THE New Science Teacher's HANDBOOK

WHAT YOU DIDN'T LEARN FROM STUDENT TEACHING

THE New Science Teacher's HANDBOOK

WHAT YOU DIDN'T LEARN FROM STUDENT TEACHING

SARAH REEVES YOUNG
MIKE ROBERTS

NSTApress
National Science Teachers Association
Arlington, Virginia

National Science Teachers Association

Claire Reinburg, Director
Andrew Cooke, Senior Editor
Amanda O'Brien, Associate Editor
Wendy Rubin, Associate Editor
Amy America, Book Acquisitions Coordinator

ART AND DESIGN
Will Thomas Jr., Director
Rashad Muhammad, Graphic Designer

PRINTING AND PRODUCTION
Colton Gigot, Senior Production Manager

NATIONAL SCIENCE TEACHERS ASSOCIATION
David L. Evans, Executive Director
David Beacom, Publisher

1840 Wilson Blvd., Arlington, VA 22201
www.nsta.org/store
For customer service inquiries, please call 800-277-5300.

Library of Congress Cataloging-in-Publication Data
Young, Sarah Reeves, author.
  The new science teacher's handbook: what you didn't learn from student teaching / by Sarah Young and Mike Roberts.
    pages cm
  Includes index.
  ISBN 978-1-936959-49-5
  1. Science teachers—United States—Life skills guides. I. Roberts, Mike, 1973- author. II. Title.
  Q147.Y685 2013
  507.1—dc23
                          2013020454

Cataloging-in-Publication Data are also available from the Library of Congress for the e-book.
e-ISBN: 978-1-938946-77-6

CONTENTS

INTRODUCTION

One of the first activities that I ask new science teachers to do in my methods course is to think of a memorable science activity from their past. Whether it's creating an exploding vinegar and baking soda volcano, dissecting a frog, or building a mousetrap car, new teachers relish their memories from school. Beyond the projects themselves, we also discuss the teachers who facilitated these learning experiences, focusing on how they inspired their students to develop a passion and curiosity for science. Once this is established, I ask my students the million-dollar question: "How are *you* going to become one of those teachers?"

Every teacher dreams of inspiring the next generation of students, and these aspirations are an unmistakable driving force for new educators. Often, this is one of the primary reasons they have chosen to enter the teaching profession. However, as with many dreams, achieving this goal can be challenging. There are many elements that have to happen before a teacher can feel confident entering that first classroom. Whether it's the courses in education pedagogy and science teaching methodology or the student teaching experience, it is a long road of hard work, learning, and preparation.

And that's where the real challenge begins.

As a new teacher, you are expected to command a classroom, implement new curriculum, and inspire future scientists. Added to this, you're supposed to do all this with the poise and confidence of a seasoned educator. It is hard to find another profession that expects near perfection in skills from those just entering the field. The first few years are difficult enough with trying to learn the schedule, understand school procedures, and mastering classroom management. Not to mention that you are also responsible for connecting with the students, establishing positive parent communications, and developing relationships with all your new colleagues. And, oh yeah, and then there is that whole "teaching" thing you're expected to do!

But as science teachers, we emulate the great scientists by never turning away from a challenge. Instead, we do what scientists do best—we use our resources.

The New Science Teacher's Handbook highlights 14 steps that you can take toward becoming a skilled classroom teacher. It identifies various challenges that new science teachers often experience, and highlights ways to prepare for and grow from them in order to become a master teacher. This book addresses areas that are often underrepresented by prevalent science methods pedagogy textbooks. By focusing on the work that takes place in the day-to-day life of a teacher, this book will prepare new teachers by reassuring them that all the unfamiliar and challenging experiences that they are going through are common among first-year teachers. In addition, this book provides resources and action plans explaining how to best prepare and take on these common classroom challenges. This text is educational for those who are preparing for a new career in teaching science, and fills the void that exists between pedagogy and the classroom.

Each chapter of *The New Science Teacher's Handbook* presents scenarios and time-tested ideas from both within and outside the classroom. The setup for each chapter is as follows:

- **The Story:** These are actual experiences that happened within either Sarah's or Mike's classroom. Any "I" statements are in reference to what happened to us individually within our classrooms. As a new teacher, it's always nice to know that someone else has had a similar struggle. These true stories demonstrate that even those who go on to write books on best practices in the classroom didn't start off as perfect educators.

- **The Moral:** What we learned from the aforementioned story. Similar to a fable, there is a moral to each story that addresses the theme of the struggle and sets the stage for moving beyond the challenge.

- **Steps for Success:** Here we present strategies to help teachers overcome situations similar to those presented in "the story." There are multiple solutions presented so teachers can choose those that work best for their specific concerns and school environment.

- **What Does Success Look Like?** This section examines how the classroom looks after implementing the "steps for success." This is the "light at the end of the tunnel" to help new teachers see that common challenges can end with positive results that benefit both teacher and students.

- **Resources:** Here we present resources to consider for additional support in organizing the classroom for those teachers who want to explore the topic in more detail.

Whether you are on your way to becoming a new science teacher or a teacher in the early years of your career, we feel confident that the ideas presented here will help you become the teacher you've always wanted to be.

ABOUT THE AUTHORS

Sarah Reeves Young

Sarah Reeves Young is the K–12 Science Specialist for the Utah State Office of Education. Young provides technical support and leadership in the development and improvement of science education in the elementary and secondary schools of the state.

Prior to moving into this role, Young was an eighth-grade physical science teacher at Rowland Hall Middle School in Salt Lake City, Utah. Aside from her work in the classroom, Young was involved in the academic realm as an adjunct professor for Westminster College, where she taught their masters students in secondary science methods, and was also the program director for the Teacher Training Institute.

Young earned a bachelor's degree in environmental, population, and organism biology from the University of Colorado at Boulder in 2004. She then pursued her master's degree in education at Lesley University in Cambridge, Massachusetts, and earned her degree in secondary science education in 2005.

Young has published in NSTA's member journal *Science Scope*, presented at regional and national conferences, and participated in several professional development institutes. She previously published a book with NSTA Press titled *Gourmet Lab: The Scientific Principles Behind Your Favorite Foods*. Young is dedicated to inspiring the next generation of scientists.

Mike Roberts

Mike Roberts currently teaches eighth-grade English at Rowland Hall Middle School in Salt Lake City, Utah, where he has educated students for the past 14 years.

Roberts earned both his bachelor's degree and masters in education from Westminster College in Salt Lake City, Utah. Aside from his work in K–12, he has also presented as a guest lecturer to new teachers at Brigham Young University and Westminster College. Roberts has also served as the program director for the Teacher Training Institute at Rowland Hall.

Additionally, Roberts has also been recognized in the community for his work engaging students with a neighborhood garden as well as senior citizen centers. He regularly brings in young adult authors to speak to students, and coordinates their visits to a network of local schools to share the experience with students and teachers from across the Salt Lake Valley.

In addition to his work with the community, Roberts has spent the past several years writing a regular column for the *English Journal*, identifying teaching ideas connected to contemporary young adult literature. Additionally, Roberts presents locally and nationally about best teaching practices for the classroom. He has won numerous awards for his innovative and engaging teaching methods including the American Stars of Teaching Award in 2008 and the National Council of Teachers of English Hoey Award for Outstanding Middle School Educator in English in 2009. Roberts has also has served on many educational committees, and currently is the assistant chair for the middle level section of the National Council of Teachers of English.

ACKNOWLEDGMENTS

Sarah Reeves Young

To my first student, Elizabeth, for letting me practice being a teacher early on, and educating me about family and friendship.

To all of the teachers who helped me get from Chapter 1 to Chapter 14, including those from Lesley University where I completed my teacher preparation. Specifically, to my coauthor Mike, for his simply stunning passion that never fails to inspire me.

Finally, to my five-year-old Jake, for asking me questions everyday. I will always grow while surrounded by your curiosity.

Mike Roberts

I want to thank my kids (Luke and Lily) for making me a better person, my grandma for teaching me to work hard, and Sarah for helping me be the best teacher I can be.

INTO THE JUNGLE

A GUIDE TO INVESTIGATING YOUR NEW SCIENCE CLASSROOM

This chapter is all about what to do when you are given your first classroom. As a science teacher, you often inherit a jungle of supplies that are organized in a way that only one person would understand … and that person is most likely retired and enjoying an umbrella drink on a beach, laughing at the new teacher who inherited the room! So how do you approach getting the classroom in order so it can be *your* space?

THE STORY

I am standing in the middle of a science classroom with a bicycle wheel. There is no bicycle to be seen, and I am at a loss. It has been over a week since I received the key to my first classroom. My initial excitement to hang up posters and move into my desk has turned to absolute horror as I look around at a classroom that is filled with … junk. There are old eyedropper bottles with remnant labels peeling off, a box of plastic-wrapped dried corncobs, a neon red wig, a box labeled "toxic" that has nothing inside, and pieces of broken crayons. And that is just the first drawer in a classroom where every shelf, closet, and cabinet is filled with various items in disarray. But after days of sifting through materials and thinking, "Was the previous teacher a hoarder?" it's the bicycle wheel that is about to push me over the edge.

As a new teacher, I thought I would be inheriting a blank slate. Instead, I have been given what I see as the equivalent to a storage locker. I was not able to meet the previous teacher in my position, and I have been given very little guidance as to what needs to be taking place in my new classroom. Looking at the supplies, it's hard to know what is trash, what is useful, and what is a biohazard. I thought I would get a head start coming in to school before the other teachers, but find myself feeling overwhelmed … and I haven't even met my students yet! I knew the numbers were daunting when I entered the field, and that one-third of teachers quit within their first three years. I just didn't think that it would be a bicycle wheel that caused my early exit before the actual school year began.

Fast forward. It's February of my first year, and I am slowly rotating on a spring stool while holding the spinning bicycle wheel to demonstrate angular momentum. The item that almost caused me to have a mental breakdown is now holding the attention of an entire class of science students. The students are amazed by translation of force, and they quickly ask to try it themselves. The excitement is palpable, and in less than a minute, all my students are out of their seats and lined up for the science experience.

"Where did you get this?" a student asks, as he anxiously waits his turn to hold the spinning bicycle wheel.

"From an amazing previous teacher who saw the science in everything," I respond out loud while thinking to myself, *Hmmmm, I wonder if this now makes me a hoarder?*

THE MORAL

Most new teachers do not have the opportunity to meet with their predecessor, and even those who do have such limited time that they often find themselves focusing on the curriculum sequence and school policies. Items such as classroom organization generally don't get passed along in the transition. It can be incredibly overwhelming to inherit a classroom where you feel more like a guest than an owner.

To add to the challenge, most science classrooms become a dumping ground for supplies. This is often because not only do science rooms have the most cabinet space, science teachers are also known for seeing the science in *everything*. This translates to teachers keeping boxes of old film canisters, *National Geographic* magazines, and four cracked aquariums because "they might be useful someday." As the new teacher in the room, it's a difficult balance between wanting a fresh start to establish yourself, but not throwing away good classroom materials that may be helpful in your hands-on lessons (that you have yet to design). Here are some helpful tips for exploring your new environment.

STEPS FOR SUCCESS

1. YOUR DESK

This tends to be the easiest place to start, since most of the materials are teacher-related supplies like staplers, writing utensils, and notepads. Evaluate what you can use, and then get rid of the rest (including empty mechanical pencils, dried out markers, broken binders, and notepads that say "From the Desk of Mr. Robinson"… unless you happen to share that last name).

Once you have narrowed the supplies down to those that are useful, evaluate them again for what you *will* use. For example, if you would never grade a paper in red ink, don't keep pens with red ink. All of the extraneous supplies can go to a student supply area, the office, or the faculty room with a sign that says, "FREE SUPPLIES from the new science teacher."

Additionally, there are often lots of old papers in a teacher's desk, and I recommend sorting them into the following piles:

- **Student work.** For student work samples, keep one example of each assignment. This can be helpful for understanding the previous course expectations and what assignments were used to achieve the course objectives. Organizing these into subject and topic area will make it easier for them to be a resource during your future curriculum planning.

- **Standards documents.** If you are teaching in the state where you received your new teacher training, you have most likely seen the science standards that help guide the courses that you have been hired to teach. Keep in mind that these documents *change* over time. You want to make sure you are using the most up-to-date version to help your classroom planning. As most of these documents are available online, old copies can be recycled unless they contain teacher notes that are useful.

  What if you have no standards? You are in luck! The *Next Generation Science Standards* were released in April 2013. This document is free from the National Academies Press (see resource section at the end of the chapter) and can be a great road map along with the associated *Framework* document of goals and student performance expectations to help you construct your classroom. Even if you have state standards, this is a document that can be instrumental for supporting new teachers.

- **Teacher plans.** Teacher lesson plans are a goldmine, and you should count your lucky stars that these were in your classroom. Some schools have old plans on file for teachers, so if you don't find these in your desk, I encourage you to ask your administration, mentor teacher, or teammates if they have any lesson plans that would be applicable to your new teaching assignment. Keep these in a separate binder, as they can be advantageous for curriculum planning. What if you can't find any lesson plans? Welcome to the club! Consider it an "opportunity" to develop all those creative ideas you came up with during your student teaching.

- **Administration.** Administrative papers such as such as bills, purchase orders, and school policies can be tricky. After all, you don't want to find out your school was audited, and that you were the teacher who threw out all of the financial files. Ask your school administration and office how far back they keep paper records and if they have specific expectations for what you need to keep. I kept all of the administrative documents in a file cabinet my first year, and I cleaned them out right before I left for my first summer. This allowed me to spend the year learning about the policies and practices of my school, as well as making sure that I had all of the required documents.

- ***Written before the age of enlightenment.*** This is a good time to find the recycling bin at your school. I was amazed to find articles that talked about microscopes as "cutting edge" technology … OK, maybe they weren't that old, but anything that predated my birth ended up being recycled.

2. BOOKSHELVES AND PAPERS

This is another great place to find curriculum and resources about your classroom and curriculum. If there is more than one of a specific book, it was most likely used as student resources. I recommend separating teacher resources from student resources, and then separating those by subject/content area. I kept all of the books in my first classroom, even those that didn't have a direct link to my subject area (as they were useful for many of the elective classes that I ended up teaching).

If there are books that you clearly don't want (romance novels, anyone?), I recommend having either the librarian take a look at them or offering them to other teachers (who may recognize Mark Twain books that should be in their classrooms). If you still have extra resources, you have just found your present for the "White Elephant" gift exchange during the holidays!

3. STUDENT SCHOOL SUPPLIES

These would be metersticks, graph paper, scissors, glue, and any other items that students will need to engage in scholarly activities. Try to get all of these items organized into a single area of your classroom (cleverly titled the "Student Supply" area). It's helpful to have student supplies in a designated space that is accessible at any time. It also helps to eliminate being asked for a ruler every two minutes. Additionally, you can keep student materials that are left in class in this area for students to use in emergencies. I recommend having this area be cleaned before students leave each day. You can do this as a community or through specific student assignment. If you set the precedent early it helps your students become more accountable for keeping the supplies neat and tidy.

4. SCIENCE LAB MATERIALS

At this point, you should have encountered resources that give you a general idea of what you will be teaching. Having a clear picture of what subjects and topics you are going to approach will help frame your last resource organization challenge: Science Lab Materials. Although there is really no foolproof method for this monumental task, here are a few thoughts to keep in mind:

- **Keep similar items together.** Beakers may be scattered throughout the room as holding vessels, but they are more useful when they are together and accessible for measuring and mixing. Having all beakers together in one place helps to establish order and allows you to use the materials that you have available to design hands-on lessons. Plastic tote boxes are ideal for organization, as they let you readily see your materials, but these can be expensive. Shoeboxes and copy paper boxes are cheaper alternatives.

- **Group items by subject area.** Even if the subject is "items that would be found in my bathroom," it gives you a better chance to establish organization. Using the topics that you uncovered in your archeological dig through the desk and bookshelves will help you find connections between supplies and what they may be useful for in the future. Take the time to label these subject areas so that you can navigate your "organized" supplies more easily during the school year.

- **All household and lab chemicals must be LABELED and STORED according to their Globally Harmonized System of Classification and Labeling of Chemicals (GHS) safety data sheets.** These sheets detail proper storage that is required. If you have trouble locating GHS sheets (which MUST be present where the chemicals are stored), you can download them from the U.S. Department of Labor site (*www.osha.gov/dsg/hazcom/index.html*). If you are worried about mystery chemicals or liquids, *do not dispose of them*. Contact your school administration and building staff about the procedures they have in place for chemical disposal. Mystery chemical disposal can be very difficult and expensive, so it's important to make sure you maintain clean and legible labels on all chemicals you purchase.

- **If there are more than 10 of any item, keep them.** You might not realize it now, but those 12 baby food jars are actually perfect for your butter-making lab that highlights physical change. If the previous teacher kept 10 or more of an item, it's probably worth setting aside to see if you need them for the school year.

- **Use the Jackson test to evaluate items.** "Does this cost more than $20?" If it does, it may be worth holding onto during your first year. It would be unfortunate to spend your entire science budget replacing the electronic balance that looked old, but still functioned. If you don't know what something costs, take a moment to look it up before setting it aside as rubbish.

- **It's OK to get rid of items.** Although it can be hard, especially since you are still unsure of the full resources you need for your year of science curriculum, you can get rid of items to make room for the materials you *know* you need. Rarely are things such as crumpled posters of the planet Pluto, old student cell models made from cracked craft clay, and a dried up bottle of purple nail polish actually useful. I would also recommend checking with your fellow teachers and teammates before getting rid of items (to see if there are things that they would like to keep). I once had a teacher claim a bag of dried corn, claiming it was "perfect for her class." This also gives the other teachers a chance to say, "You know the last teacher used these for _____." If this happens, it gives you one last chance to rescue the items before they are taken away for good.

5. CLASSROOM INTERIOR DESIGN

Consider how you might rearrange your classroom furniture. Learning the art of *feng shui* for the classroom can be helpful to facilitate different learning environments such as group work, presentations, discussions, stations, and partner activities. It's helpful to try out these orientations prior to students arriving so you know what options are possible (and *not* possible) for all the desks in your room. It's also a good time to find out how many students you will have, so you can get on the list for needing additional chairs or desks (if need be) before the first day of school.

6. PERSONAL ITEMS

Finally, add your own personal touches to the classroom. Students are interested to learn more about you, so it's OK to include a few personal items (such as your favorite book, a poster of your favorite band, or a magnet with your favorite quote) here and there. I taught with a teacher who had an old school photo of himself from grade school on his bulletin board that sparked great conversation in his classroom. Having personal items will also help you feel more ownership over your new space. But don't go overboard! You don't need your bulletin board to be your new Facebook or Pinterest page, but chose a couple of items that are important to you and low in monetary value that you would be comfortable sharing with your students and fellow colleagues.

7. BLANK CANVAS?

Some new teachers do inherit a blank canvas, and look at those teachers who have too many supplies like the kids who got too much ice cream at lunch: not with a lot of sympathy when you have too much. Besides the resources referenced in this chapter for constructing a classroom, visit Chapter 3 (budgeting) and Chapter 10 (professional development) to learn how to provide your classroom with supplies.

WHAT DOES SUCCESS LOOK LIKE?

By taking time to organize your classroom at the start of the school year, you will find yourself much more prepared for tackling the other challenges of being a new teacher (including learning school schedules, student names, developing relationships with colleagues, and so on). Organizing the space helps you better understand the curriculum, learn about what supplies are available to you, and gives you ownership over your new space. This translates into confidence and better planning for the outset of the school year.

Take my experience with the bicycle wheel. By the time February had hit and we had reached Physics, I was well beyond using my lessons from my time as a student teacher, and I was often searching for activities and events the day before we discussed the topic. I found a great demonstration for angular momentum, and low and behold, it required a bicycle wheel! And while that wasn't an item that I was going to be able to pick up at the local grocery store, I knew that I had it in my classroom because I had taken the time to organize the space over the summer. I was able to walk into the classroom and find the materials I needed to give a great demonstration that would get my students excited about momentum.

The other major benefit that came from the initial cleanout and organization was the sense of ownership that came with the process. This classroom was now mine, and a milestone in fulfilling my dream of becoming a teacher. There is nothing quite as satisfying as hanging up your first science poster in the room where you will have the opportunity to influence and support your students' appreciation for science. And once it became *my* classroom, I felt more confident and comfortable creating and leading the learning community. Whether it was knowing where I could go to get a larger graduated cylinder, or having my favorite science book (*A Short History of Nearly Everything* by Bill Bryson) available on my shelf, knowing my classroom and supplies allowed me to construct the classroom atmosphere that I had been dreaming of since I first started on the road toward becoming a science teacher.

RESOURCES FOR MORE INFORMATION

Achieve Inc. 2013. *Next generation science standards.* www.nextgenscience.org/next-generation-science-standards

Next Generation Science Standards identify the science all K–12 students should know. These new standards are based on the National Research Council's *A Framework for K–12 Science Education.* The National Research Council, the National Science Teachers Association, the American Association for the Advancement of Science, and Achieve partnered to create standards through a collaborative state-led process. The standards are rich in content and practice and arranged in a coherent manner across disciplines and grades to provide all students an internationally benchmarked science education.

National Research Council (NRC). 2012. *A framework for K–12 science education: Practices, crosscutting concepts, and core ideas.* Washington, DC: National Academies Press.

Science, engineering, and technology permeate nearly every facet of modern life and hold the key to solving many of humanity's most pressing current and future challenges. The United States' position in the global economy is declining, in part because U.S. workers lack fundamental knowledge in these fields. To address the critical issues of U.S. competitiveness and to better prepare the workforce, *A Framework for K–12 Science Education* proposes a new approach to K–12 science education that will capture students' interest and provide them with the necessary foundational knowledge in the field.

A Framework for K–12 Science Education outlines a broad set of expectations for students in science and engineering. These expectations will inform the development of new standards for K–12 science education and, subsequently, revisions to curriculum, instruction, assessment, and professional development for educators. This book identifies three dimensions that convey the core ideas and practices around which science and engineering education in these grades should be built. These three dimensions are: crosscutting concepts that unify the study of science through their common application across science and engineering; scientific and engineering practices; and disciplinary core ideas in the physical sciences, life sciences, and Earth and space sciences and for engineering, technology, and the applications of science. The overarching goal is for all high school graduates to have sufficient knowledge of science and engineering to engage in public discussions on science-related issues, be careful consumers of scientific and technical information, and enter the careers of their choice.

A Framework for K–12 Science Education is the first step in a process that can inform state-level decisions and achieve a research-grounded basis for improving science instruction and learning across the country. The book will guide standards developers, teachers, curriculum designers, assessment developers, state and district science administrators, and educators who teach science in informal environments.

Motz, L. L., J. T. Biehle, and S. S. West. 2007. *NSTA guide to planning school science facilities, second edition.* Arlington, VA: NSTA Press.

Science-learning spaces are different from general-purpose classrooms. So if your school is planning to build or renovate, you need the fully updated *NSTA Guide to Planning School Science Facilities*. It's the definitive resource for every K–12 school that seeks safe, effective science space without costly, time-consuming mistakes.

New to the revised edition is a chapter on "green" schools, including how to think outside the traditional walls and use the entire grounds to encourage environmental responsibility in students. The guide also provides essential up-to-date coverage such as:

- Practical information on laboratory and general room design, budget priorities, space considerations, and furnishings.
- Stages of the planning process for new and renovated science facilities.
- Current trends and future directions in science education and safety, accessibility, and legal guidelines.
- Detailed appendices about equipment-needs planning, classroom dimensions, and new safety research, plus an updated science facilities audit.

NSTA Guide to Planning School Science Facilities will help science teachers, district coordinators, school administrators, boards of education, and schoolhouse architects develop science facilities that will serve students for years to come.

Unger, M. S. 2011. *Organized teacher, happy classroom: A lesson plan for managing your time, space, and materials.* Blue Ash, OH: Betterway Home.

Organized Teacher, Happy Classroom will give readers a fresh perspective on organizing that will promote productivity and efficiency, allowing them to focus more on student achievement and worry less about keeping their classroom materials in order. Readers will find specific help with purging their unused materials and papers, creating filing systems and managing daily routines. Summative checklists at the end of each chapter will help readers apply key principles as they organize.

Heyck-Merlin, M. 2012. *The together teacher: Plan ahead, get organized, and save time!* San Francisco: Jossey-Bass.

Maia Heyck-Merlin helps teachers build the habits, customize the tools, and create space to become a "Together Teacher." This practical resource shows teachers how to be effective and have a life! Author and educator Maia Heyck-Merlin explores the key habits of Together Teachers—how they plan ahead, organize work and their classrooms, and how they spend their limited free time. The end goal is always strong outcomes for their students.

So what does Together, or Together Enough, look like? To some teachers it might mean neat filing systems. To others it might mean using time efficiently to get more done in fewer minutes. Regardless, Together Teachers all rely on the same skills. In six parts, the book clearly lays out these essential skills. Heyck-Merlin walks the reader through how to establish simple yet successful organizational systems. There are concrete steps that every teacher can implement to achieve greater stability and success in their classrooms and in their lives.

Segura, H. 2011. *Less stress for teachers: More time and an organized classroom.* San Antonio, TX: Hacienda Oaks Press.

This organizing book for educators will teach you:

- How to save a minimum of 90 hours per school year
- The five most critical areas to control during your day
- How to manage e-mail, paperwork, lesson plans, and other tasks
- How to set up your classroom in the most efficient way
- How to lower your stress level during the school day

Implementing the ideas in this book will save you a minimum of 30 minutes per day, which is 2.5 hours per school week, and 90 hours per school year! That's the equivalent of 11+ school days!

Less Stress for Teachers: More Time and An Organized Classroom addresses the thinking behind how to overcome "the overwhelm" that teachers feel on a daily basis. Teachers are bombarded with hundreds of tasks and decision-making situations per day, but they are often not given the tools to cope with all of that.

Learn how to become an organized teacher with an organized classroom. Use this as a first-year teacher's survival guide or a veteran teacher's journey to achieving classroom organization and less stressful school days.

WORKING WITH A MENTOR

HOW TO APPROACH THE RELATIONSHIP

This chapter is about how to go about creating a working relationship with a mentor teacher. Most new teachers are assigned mentor teachers, who get the position based on the number of years they have worked in a school, not necessarily their ability to coach a young teacher. The chapter looks to address how new science teachers can engage their mentor in a way that is constructive and beneficial to the first years of teaching.

THE STORY

It's early November, and I am surviving my first year of teaching but really feel like I am on my own island. I spend almost all of my free time in my classroom cleaning, setting up, planning, and grading. I rarely even go down to the faculty lounge to eat lunch with the other teachers, since nutritional sustenance is playing a secondary role to my classroom priorities. Since I spend so much time trying to be the great teacher I learned about in my teacher preparation program, I don't really interact with my colleagues much, as they all seem very calm and put together. After all, the last thing I want them to know is that I am five minutes away from not having a clue about what I am doing!

But for today's lesson, I feel confident in its success, as I have students actively involved in an inquiry-experiment studying states of matter. The key word here is "active." Although the inquiry-based activity was designed to have students rotate through stations, my classroom is in chaos. The students aren't taking the time to read any of the directions on the lab that I handed them at the beginning of class. Instead, they are moving from area to area, taking materials with them and scattering the items that had taken me 45 minutes to set up. And as I am trying to clean up the lake of water that is forming at the "liquid" station, I hear from across the room that the salt test tube is missing from the solids station. I turn around just in time to watch it being thrown from the gas station, where I have students spraying air freshener like it's the middle school gym locker room.

"Do you have any rulers?" one of my students calmly asks amidst the bedlam.

"What? Rulers? Aren't they in the student supply cabinet?" I say as I run to see if I can find some sort of measuring instrument. The student supply cabinet might as well be empty based of the incredible disarray. The likelihood of finding a ruler is about the same as finding a needle in a haystack. But never one to let adversity stop me, I resolve to *run* across the hall to ask the math teacher if she has any rulers that I can borrow. I take one last look at my students and yell "LAB SAFETY RULES ARE IN EFFECT!" as I duck out to find measuring instruments.

Recognizing that I have just failed teacher supervision 101 by leaving my room unattended, I race to the math room with the intent of returning to my classroom in less than 20 seconds.

But when I walk through my mentor's door, I am stopped dead in my tracks.

The math classroom appears to be a parallel universe, where students are sitting quietly in pairs working on solving linear equations and contemplating alternative solutions to build their comprehension. The learning is palpable, as I see the math teacher, Jocelyn, helping a pair of students. She sees me and calmly walks over to the doorway, where I am still in stunned silence over the engagement and structure of her room. I take the rulers (which were all neatly in her supply cabinet and labeled with her name) and return to my room to find a sponge fight in full effect.

I went home that night feeling defeated. What was wrong with me? I had passed my classroom management class with flying colors, writing an outstanding paper on the importance of lab safety. The sponge fight, however, begged to call my expertise into question. I was trying incredibly hard to engage the students in inquiry-based learning, but my classroom had devolved into anarchy. There was clearly a lack of structure and although I was trying my best, my best was obviously not good enough.

After having a good cry over the contrast between Jocelyn and myself, I recognized that I had two options:

1. Continue down this path of destruction and hope that my academic excellence from my graduate school studies was just delayed and would kick in by December.

   Or …

2. I could do what I always tell my students to do when confronted with a problem: SEEK HELP. I always encouraged them to engage their peers or me when they were struggling with a science concept or skill. Why should it be any different for me?

Although it was in contrast to the all-knowing teacher that I so desperately wanted to be, I went in early the next morning and approached the math teacher.

"I'm struggling with classroom management," I said as quickly as I could, knowing that any moment I might lose my courage to admit my faults.

"Oh good. I'm glad you've noticed." She smiled back at me.

THE MORAL

There are some people who are just born to be educators. Their transition from student teacher to classroom teacher is effortless, and students and colleagues alike would never guess that they were new teachers.

I, along with the other 99% of new educators, was not one of those teachers.

Apparently, my secret struggle with classroom management wasn't so secret. It turns out that many of my colleagues knew, but they had hesitated to reach out to me because they didn't want to overstep their boundaries. Instead, they were waiting for me to reach out and ask for help.

This was a huge realization for me. During my student teaching, my mentor frequently asked how I was doing, if I needed any help, and offered suggestions to improve my lessons. Because of this, I assumed that, as a new teacher, my mentor and colleagues would do the same. But it's incredibly important to understand that as a new teacher (versus being a student-teacher), you *must* reach out to your mentor and colleagues and ask for help when needed. This lets your mentor know that you are not only open to feedback, but also that you are seeking support in ways beyond the monthly "touch base" meeting.

STEPS FOR SUCCESS

To establish a supportive mentor relationship, try some of the following ideas:

1. SHARE YOUR STRENGTHS AND WEAKNESSES.

Identify one teaching skill you are doing well, as well as one where you are struggling, during your monthly meetings. This gives you an opportunity to get accolades for your successes, while at the same time seeking advice and guidance in areas where you are having difficulty. Looking at your instruction in this manner allows you to understand your growth as a teacher through a constructive conversation rather than via a subjective evaluation.

2. ASK YOUR MENTOR TO OBSERVE YOUR CLASS WITH A SPECIFIC FOCUS.

Highlight one aspect that you would like their feedback on. Some examples include:

- "I'm having trouble with a specific student, can you observe and help me understand what is contributing to the problem?"
- "Can you observe my movement through the classroom and give me pointers on how to more effectively be away from the front of the classroom?"

- "I am working on incorporating a range of questions and depth of knowledge in my discussions. Can you observe to see how many of each we address in my class today?"
- "When calling on students, I want to make sure that I am giving everyone a chance. Would you observe and record how many times each student is called upon in class today?"

Classroom observations can be very vague if there isn't a targeted goal in mind. Take leadership in these opportunities by giving direct criteria for what you would like for your mentor to look for in your lesson. This allows you to gain both positive and constructive feedback, while at the same time, it establishes a more meaningful observation for both you and your mentor.

3. ASK TO OBSERVE YOUR MENTOR'S CLASSROOM WITH A SPECIFIC FOCUS.

This can be incredibly helpful for learning about the various teaching approaches that you could use in your classroom. For example, if you are struggling with transitions between activities, watch how your mentor teacher handles these situations. Observing one specific element of a class will not only help you become a better teacher, it can also help you gain a stronger understanding of what students are experiencing in other courses. Look for approaches that are successful so you can find ways to adjust them to work in your classroom community.

4. ASK YOUR MENTOR FOR RECOMMENDATIONS OF OTHER TEACHERS TO VISIT FOR OBSERVATIONS AND SUPPORT.

Use their knowledge of the school community to help you build your network of support from a *single* mentor to a *cohort* of teachers. The more you communicate with others, the more ideas you will have for solving the problems you encounter as a new teacher. If your school uses a Professional Learning Community model (PLC), those groups and meetings can be a great place to start reaching out to other educators for suggestions, support, and observation.

5. ATTEND PROFESSIONAL DEVELOPMENT WITH YOUR MENTOR.

These are great opportunities for not only learning together, but also often lead to collaborative projects between the two of you. This translates itself into a stronger relationship throughout your mentorship (and beyond), including the (many) times when you will be in need of support.

It's important to note that not all mentors are officially "assigned" the job. Many times, you will find support from teachers beyond the traditional mentor relationship (especially if your school doesn't have a formal mentor program). Working with your colleagues is essential to your development as a teacher, and it should translate into feeling like a supported member of your school community. There are many research studies that speak to the value of a mentor relationship for retaining teachers in the profession beyond their first three years, and the best advice I can offer is to find the best teachers in your school and ask them for help.

WHAT DOES SUCCESS LOOK LIKE?

"One of the easiest ways to have better classroom management is to have a daily schedule on the board. It helps students feel more secure about the expectations for the day. It provides motivation for students who are dragging their feet, and for those who finish early, it lets them know what they can move on to without distracting their peers."

Genius. The math teacher was a genius.

Aside from the schedule suggestion, Jocelyn helped me understand other ways to improve my classroom. For example, she explained how giving students the instructions *before* the class period would make them more productive *during* the class time.

She also brought in our English teacher, Mike, to get his ideas and feedback as well. Mike was the resident rock star teacher, and I had been too intimidated to approach him because of the rave reports I regularly heard from students and parents about his class. From observing his class, I was able to learn the concept of having the students tell me the directions before beginning an activity, an idea that I easily could incorporate into my class.

Long story short: After realizing that my struggle with classroom management was well known by my colleagues, I actually felt relieved. Now I didn't need to pretend to have mastery in an area where I really needed advice, and that opened up the channels for increased communication and support. I suddenly went from judging myself and competing with these educators to seeing them as resources for inspiration and learning.

I can say with all honesty that it was the network of support I received from the other teachers that helped me through my first year. If I had chosen to go it alone, I would have never survived. As a new teacher, take the lead and ask your mentor and fellow teachers for advice and support. It will save you from your own sponge battle.

RESOURCES FOR MORE INFORMATION

Zubrowski, B., V. Troen, and M. Pasquale. 2008. *Making science mentors: A 10-session guide for middle grades*. Arlington, VA: NSTA Press.

Many peer-mentoring guides *claim* to be unique. This one is. It trains middle school science teachers to be effective mentors for other middle school teachers—and does so using a long-term, inquiry-based approach to teaching and learning how to be a more effective science educator. Developed under a grant from the National Science Foundation, the principles of this guide's procedures and materials were field-tested with 50 mentors and new teachers in a variety of middle schools.

Making Science Mentors comes with everything you need to set up and run a comprehensive program:

- 10 session-by-session lesson plans
- A planning and observation protocol to guide mentor-mentee interaction, both in conferences and during classroom observation
- Handouts and citations (on paper and on CD) for use in preparing mentors
- Video clips on DVD that show middle grades science classrooms and teacher mentoring

Making Science Mentors is ideal if you have state, district, or school-based teacher preparation programs and don't want to start from scratch when setting up a research-based mentoring system. As the authors promise, "Using this guide, you will master inquiry-based mentoring and share it with mentors, who, in turn, will share it with their mentees, all in the name of improving the classroom experiences—and achievement—of students."

Rhoton, J., and P. Bowers. 2003. *Science teacher retention: Mentoring and renewal*. Arlington, VA: NSTA Press.

"Some 40% of all new science teachers leave the profession within five years, and too many science teachers are wedded to their textbooks and the routines they acquired during their collegiate years." What can be done to retain new science teachers and reinvigorate more experienced science teachers? Allow *Science Teacher Retention: Mentoring and Renewal* to "mentor" you as you reach toward this lofty but attainable goal.

For this book, Jack Rhoton and Patricia Bowers assembled some of the country's most noted science educators and asked them to offer ideas to resolve the problems of science teacher retention and renewal. Their suggestions are designed to keep the brightest and most motivated new teachers in the profession and help all science teachers to continue to learn and to treat their own profession like science itself—that is, by basing it on questions, suggesting answers, and using their interests and abilities to test the validity of these answers.

Mundry, S., and K. E. Stiles. 2009. *Professional learning communities for science teaching: Lessons from research and practice.* Arlington, VA: NSTA Press.

What would it take to move your school or district closer toward a culture that supports and sustains professional learning communities (PLCs)? This thought-provoking collection of stories detailing seven successful approaches to developing PLCs will inspire you to find answers to this question and others. From one school taking the initiative to create its own collaborative environment to a network of 500 universities and schools, the accounts explain the advantages of PLCs for teachers and their students. The volume editors begin with the argument that in a PLC environment, teachers receive continuous professional development, therefore improving their teaching skills to the benefit of student learning. Later chapters recount the origins of schools as professional learning communities, define the characteristics of professional learning communities, and review research on the subject.

Teachers and school administrators will particularly appreciate the reflection questions at the end of each chapter that encourage you to relate your learning to the chapter's specific story. An appendix provides even more resources about developing PLCs.

Professional learning communities have value in more than just school contexts, as supported by a growing body of research. In sharing the experiences of educators who have embraced the principles of PLCs and integrated them into schools and district, regional, and state initiatives, the editors hope to inspire new contributions to the transformation of science education.

Breaux, A. L., and L. S. Brandt. 2011. *101 "answers" for new teachers and their mentors: Effective teaching tips for daily classroom use.* Larchmont, NY: Eye on Education.

The second edition of this bestselling title features brand-new strategies plus illustrations! Make sure your new teachers are ready for the realities of the classroom. Be confident that their mentors are focused and effective. Organized so new teachers can read it by themselves, this book can also be studied collaboratively with veteran teachers who have been selected to mentor them. Addressing the questions and struggles of all new teachers—with simple solutions—this book

- generates instant impact on teacher effectiveness;
- promotes communication between new teachers and their mentors; and
- offers strategies for any teacher looking to become more effective.

Hicks, C. D., N. A. Glasgow, and S. J. McNary. 2004. *What successful mentors do: 81 research-based strategies for new teacher induction, training, and support.* Thousand Oaks, CA: Corwin.

Provides a wide range of practical suggestions for mentors that are based on current research and that can be "harvested" whenever needed. The "Precautions and Possible Pitfalls" sections serve as a welcome safety net, helping mentors to proactively examine and strategize solutions for anticipated challenges. The 'Sources' sections are particularly helpful in offering additional readings for those mentors who are interested in going deeper into a topic. *What Successful Mentors Do* is easy to use, linked to best practices, and is certain to be an invaluable resource for new and returning mentors.

Lieberman, A., S. Hanson, and J. Gless. 2011. *Mentoring teachers: Navigating the real-world tensions.* San Francisco, CA: Jossey-Bass.

A useful guide for teacher mentors as they face new and difficult challenges in their work. New teachers often struggle to apply their knowledge in real-world settings, and the idea of mentoring these teachers during their first years in the classroom has captured the imagination of schools all over the world. Drawn from the experiences over the last twenty years of the New Teacher Center, the book illuminates the subtleties and struggles of becoming an excellent, effective mentor. The book discusses the five big tensions of mentoring: developing a new identity, developing trusting relationships, accelerating teacher growth, mentoring in challenging contexts, and learning leadership skills.

- Describes in-depth the most common challenges of the mentor role
- A wonderful guide for both new and veteran mentors
- Includes engaging firsthand narratives written by mentors working in a variety of settings

This book is from the New Teacher Center, an organization whose highly respected mentor-training model has served over 50,000 teachers nationwide. The New Teacher Center is dedicated to improving student learning by accelerating the effectiveness of teachers and school leaders through comprehensive mentoring and professional development programs.

LIFE ON A BUDGET

HOW to USE YOUR MONEY FOR CLASSROOM SUPPLIES WISELY

T his chapter details how new teachers can make the best use of their science classroom funds, no matter how small the budget. The chapter will outline how a teacher can spend wisely throughout the year, use community resources for free gear, and know what items are worth investing in for use in the classroom beyond a single school year.

THE STORY

"Do you think it would be OK if I went and bought some salt and sugar?" I asked my principal in a quivering voice.

"Are you planning on making cookies?" he responds with a laugh.

It's my third week as a teacher, and I am on the precipice of having my first month of teaching science under my belt. I have managed to learn all of my students' names, half of the faculty names, and even what times I get to take my bathroom breaks. My intensive summer planning is paying off, and I am ready for my first investigation with chemicals using sodium chloride and sucrose (also known as salt and sugar). However, I have run into a snag—I have no salt or no sugar. Although I have contemplated purchasing them myself, I realize that my starting teacher salary is not going to be able to financially support my plans for inquiry-based lessons for the full school year. As a result, I find myself standing in front of my principal asking whether or not I have available funds for such expenditures.

"Of course you can buy sugar and salt," he answers as I release a huge sigh of relief. As a student teacher, I had heard of schools where they ask teachers to buy their own paper, and I was grateful that my school was going to allocate funding for my supplies. "Just make sure you don't spend it all too quickly. You have a limited budget, you know."

A limited budget?

Was he telling me that my science class was going to be run on the same ramen noodle budget that I lived on in college? I mean, how would I know what to spend, where to save, and how to stretch my budget to support good science education?

My savers mentality from graduate school kicks in, and I head straight to the faculty lounge to track down some sugar packets I had seen by the coffee maker…

THE MORAL

As a new teacher, budgeting was not at the top of my list of immediate needs to cope during the day-to-day science teaching experience. I could barely budget my *own* finances, let alone trying to figure out the financial planning for my science course. I lived in constant fear of running out of money my first year and, therefore, often asked my students to learn from a demonstration rather than the hands-on and interactive lesson I had originally planned. I scrimped on supplies to make sure that I had an "emergency fund," and my students' experiences within my class were hampered by my need to save.

At the end of the year, after discovering that I had a large surplus in my budget, I realized the severity of my poor judgment. My cheapskate mentality ultimately backfired when I was notified that I would not be able to roll over the funds into the next school year. Although I was able to buy some replacement glassware, I had wished that I had been more active about spending my budget on the students during the school year.

STEPS FOR SUCCESS

In an effort to avoid a similar predicament, I recommend that you take the following financial planning steps as a new teacher:

1. THE BEST PLANNING STARTS WITH KNOWLEDGE.

Ask your colleagues and/or administration if you have a science budget. If you do, GREAT! You need to ask how much, what the process is for spending those funds, and be aware of any rules about yearly carryover limits. If you do not have a science budget, it at least gives you a context for planning your future lessons. (There are more tips below for saving money on science equipment that can help your bottom line.)

2. MANY SCIENCE TEACHERS SHARE A BUDGET WITH OTHER TEACHERS AND/OR DISCIPLINES.

If this is the case, contact those teachers to see if they have specific policies and allocations for the division of assets. It's important to ask these questions early in the school year. Money can be spent quickly, and you could lose out on potential supplies if you don't join in the fiscal planning process. Also, don't hesitate to ask for what you anticipate needing to support good science education in your course. You can either base the request on the number of students you have and a set lab fee per student, or you can outline the lab experiences that you have planned and estimate costs. Either way, it's best to approach the group planning with estimates and data to support and justify your requests for money.

3. BREAK SCIENCE LAB SUPPLIES INTO TWO SEPARATE CATEGORIES: EQUIPMENT AND CONSUMABLES.

a. ***Equipment.*** Equipment supplies are going to be items that can be used repeatedly for multiple science experiences, including thermometers, balances, microscopes, and beakers. These will often be some of the most expensive items that you have in your science classroom, and since the cost can be high, be sure to consider how often you will need the equipment. If it's only once or twice, is there a way you can borrow from another teacher in your school or district, or split the cost and share the resource?

I recommend waiting until the end of your first year to make major equipment purchases. This allows you to replace any materials that were broken during the school year, and it also gives you a better perspective for what items are truly necessary versus those that can be substituted or borrowed. Equipment can also be a great reason to write a grant request, since the supplies can be used year after year to impact a larger number of students.

b. ***Consumables.*** Consumable supplies are those that are used once and then need to be replaced. These items include chemicals, paper products, glue, and perishable items. While most of these items are going to be cheaper than your equipment, when you purchase them in large volumes to meet the needs of your entire class, the costs can add up quickly. When looking at consumables to purchase, consider the following:

- Can I lessen the recommended volume or amount used by each student? This could be reducing the suggested amounts in half or using student lab groups (as opposed to each student doing his or her own). Doing an activity as a demonstration or as part of a station are two other methods to help reduce the amount of necessary supplies.

- Is there a more sustainable way to achieve the same outcome? Initially, I used large quantities of craft sticks for engineering projects until I realized that toothpicks were much cheaper and provided the same experience on a smaller scale. Drinking straws were another great option, as they were reusable, even after being covered in glue. Look for cheap or reusable alternatives to suggested supplies.

- Can I ask my students to provide any of these items? This is dependent on your community, but I found that even making use of school-issued items such as pencil bags to act as weights cut down on the costs. Keep in mind you don't want to require students to bring in materials that are costly, since not all students have access to the same resources. Try to think of items they could save or gather from the lunchroom to eliminate the need to purchase materials.

- Set aside a percentage of your budget for consumables. This will leave you money for larger equipment purchases at the end of the school year, along with giving you a starting point to evaluate the sustainability of your classroom practices. Keep in mind, the less you spend on consumables, the more you can allocate toward reusable science supplies.

4. SHOP AROUND FOR A GOOD DEAL.

There are many places where you can buy supplies for a science classroom, and it's important to examine all of your options. For example, when I was looking to purchase a light meter to measure the intensity of a light source, I found that it was much cheaper to purchase a model from a photographer supply website than it was to buy the same version from a science supply source. On the other hand, I found that I could purchase experiment sets from this same science supply company for much less than I could if I purchased each chemical individually. Many times, if you buy in bulk, you can get an even better deal, so plan ahead and coordinate with other teachers to help get the best possible price.

5. SCIENCE IS THE STUDY OF THE WORLD AROUND US, AND THAT WORLD CAN BE A GREAT PLACE TO FIND FREE OR LOW-COST SCIENCE EQUIPMENT.

I once met a teacher at a conference who made a "science scavenger hunt" list filled with items that students could find in their houses, neighborhoods, or through contacts with science professionals. They had a contest amongst each class to see who could find the most items. By the end of the contest, the teacher had enough supplies to last him for over four years!

6. SHARING IS CARING.

Don't hesitate to ask other teachers if they would be willing to share supplies with you. The key here is to ask ahead of time and to return the supplies in the same condition that you borrowed them. Additionally, in an effort to follow the golden rule, be prepared to share your supplies as well. It's a good idea to have your important supplies marked with your last name so that they make it back to you (thus saving your future budgets).

7. IT'S OK TO PLAY THE SYMPATHY CARD BY TELLING PEOPLE YOU ARE A TEACHER IN THE INTEREST OF GETTING SUPPLIES FOR SCIENCE.

I was at a restaurant where they used multicolored straws with multiple bend points. I asked the server where I could get some, sharing that they would be great for a project that I was doing with my students. At the end of the meal, the server gave me a package of 150 straws to take back to my classroom. Many people support education, and some retail locations even offer discounts for teachers. I wouldn't advocate wearing a sandwich board asking for free science supplies everywhere you go, but sharing you're a teacher can help to stretch your budget and improve your access to supplies.

WHAT DOES SUCCESS LOOK LIKE?

Knowing your budget can help you use your resources more wisely. Hands-on science experiences, the kinds that engage students and help lead them to discover new ideas, requires materials. And you can't design and plan those experiences until you know the resources that are available to you. Understanding your budget also helps you to become more thoughtful about saving reusable supplies, as well as helping to find creative ways to get supplies that seem beyond your means.

Success also means having food on the table when you go home from the classroom. Teachers are altruistic in nature, and we are often the first to go without if it means supporting student learning in the classroom. The thousands of teachers who spend money out of their own pockets each year to buy supplies for their classrooms demonstrate this. But even with your own money, you have limits to what you can provide. You want to make sure that you are using your allocated resources wisely so that you can offer your students the best education available without finding yourself behind in your rent.

The following year, when it was time for the salt and sugar experiment again, I went to the store and bought an entire container of salt AND a bag of sugar. It wasn't my annual cost-of-living pay raise that made this splurge possible, but instead, it was the knowledge and understanding of what of my class budget was that accomplished this. Because of my new financial awareness, both student engagement and the overall science education of my class were improved. It also helped me to enjoy my pancakes and to stop eyeing the saltshakers at the local IHOP restaurant.

RESOURCES FOR MORE INFORMATION

Tax Benefit for Teachers Who Purchase Supplies: *www.irs.gov/taxtopics/tc458.html*

If you are an eligible educator, you can deduct up to $250 ($500 if married filing joint and both spouses are educators, but not more than $250 each) of any unreimbursed expenses [otherwise deductible as a trade or business expense] you paid or incurred for books, supplies, computer equipment (including related software and services), other equipment, and supplementary materials that you use in the classroom. For courses in health and physical education, expenses for supplies are qualified expenses only if they are related to athletics. This deduction is for expenses paid or incurred during the tax year. The deduction is claimed on either line 23 of Form 1040 (PDF) or line 16 of Form 1040A (PDF).

COST-SAVING RESOURCES

Rich, S. 2012. *Bringing outdoor science in: Thrifty classroom lessons.* Arlington, VA: NSTA Press.

When it's just not possible to take students out to explore the natural world, bring the natural world to the classroom. Clearly organized and easy to use, this helpful guide contains more than 50 science lessons in six units: Greening the School, Insects, Plants, Rocks and Soils, Water, and In the Sky. All lessons include objectives, materials lists, procedures, reproducible data sheets, and ideas for adapting to different grade levels, discussion questions, and next steps. Almost all the needed materials are inexpensive or even free (such as leaves and rocks), and if you do get the chance to venture outside, the lessons will work there, too. By using Steve Rich's follow-up to his popular book *Outdoor Science: A Practical Guide*, you can introduce students to everything from bug zoos to the Sun and stars without ever needing to pull on a jacket.

Horton, M. 2009. *Take-home physics: 65 high-impact, low-cost labs.* Arlington, VA: NSTA Press.

Take-Home Physics is an excellent resource for high school physics teachers who want to devote more classroom time to complex concepts while challenging their students with hands-on homework assignments. This volume presents 65 take-home physics labs that use ordinary household items or other inexpensive materials to tackle motion and kinematics; forces and energy; waves, sound, and light; and electricity and magnetism. The result: Students learn background knowledge, reinforce basic process skills, practice discovery, and bridge classroom learning with real-world application—all while getting excited about homework. Teachers can also integrate science and literacy by requiring the use of lab notebooks with formal write-ups. Materials lists and safety notes, as well as both student activity pages and teacher notes are included.

Froschauer. L. 2010. *The frugal science teacher, 6–9: Strategies and activities.* Arlington, VA: NSTA Press.

Teachers of all grades and disciplines often dip into their own wallets to outfit their classrooms with materials and supplies that school and district budgets can't—or won't—cover. Science teachers tend to find themselves supplementing their shrinking funds with even greater frequency.

This collection of essays, carefully selected by former NSTA president and current *Science and Children* editor Linda Froschauer, outlines creative and inexpensive ways for sixth- through ninth-grade science teachers to keep their expenses to a minimum in five categories:

- Student-Created Constructions
- Teacher-Created Constructions and Repurposed Materials
- Teaching Strategies That Maximize the Budget
- Instructional Lessons That Maximize the Budget
- Funds and Materials

Chapters provide inexpensive alternatives to costly classroom projects, offer re-imagined uses for items teachers already have at home or school, and suggest new and untapped resources for materials. Even more important than offering ideas for frugality, the activities and strategies—such as "Wiffle Ball Physics," "Geology on a Sand Budget," "Forensics on a Shoestring Budget," and "Ever Fly a Tetrahedron?"—enhance teachers' abilities to develop their students' conceptual understanding. A comprehensive list of the many free resources available from the National Science Teachers Association is also included.

"By following the recommendations found in this book," writes Froschauer, a retired classroom teacher of 35 years, "you will find creative ways to keep expenses down and stretch your funds while building student understanding."

Horton, M. 2011. *Take-home chemistry: 50 low-cost activities to extend classroom learning.* Arlington, VA: NSTA Press.

For high school science teachers, homeschoolers, science coordinators, and informal science educators, this collection of 50 inquiry-based labs provides hands-on ways for students to learn science at home—safely. Author Michael Horton promises that students who conduct the labs in *Take-Home Chemistry* as supplements to classroom instruction will enhance higher-level thinking, improve process skills, and raise high-stakes test scores. Many of the exercises involve skills such as measuring, graphing, calculating, and extrapolating graphs, and cover topics such as moles, chromatography, chemical reactions, and titration. Each lab includes both a student page and a teacher page and provides an objective, a purpose, a materials list, notes, and postlab questions, making *Take-Home Chemistry* a useful tool for improving how students learn chemistry.

Cunningham, J., and N. Herr. 1994. *Hands-on physics activities with real-life applications: Easy-to-use labs and demonstrations for grades 8–12.* San Francisco, CA: Jossey-Bass.

This comprehensive collection of nearly 200 investigations, demonstrations, mini-labs, and other activities uses everyday examples to make physics concepts easy to understand. For quick access, materials are organized into eight units covering Measurement, Motion, Force, Pressure, Energy and Momentum, Waves, Light, and Electromagnetism. Each lesson contains an introduction with common knowledge examples, reproducible pages for students, a "To the Teacher" information section, and a listing of additional applications to which students can relate. More than 300 illustrations add interest and supplement instruction.

STARTING CLASS THE RIGHT WAY

STARTER ACTIVITIES

A successful class period is often made within the first few minutes, and this chapter will detail effective and engaging science-based starter activities. These 10-minute teaching ideas, including "Admit Slip, Please!," "Inner/Outer Circle," "The Top Ten," and "Circle Time," are easily adapted to fit into any teaching unit.

THE STORY

I was weeks away from my very first summer vacation as a teacher. The school year had been long and difficult, but I was feeling good because I had consistently turned out units that my students found engaging. I had used all the tips and tricks that my education classes had taught me (differentiation, multiple intelligences, using pop culture to make connections, and so on), and I pretty much had things dialed in at this point in the year.

Or so I thought.

After filling the board with notes in a rainbow of colors, my first period began to filter in one by one. Kristen, one of my better students at the time, casually walked in the room and, upon seeing the notes on the board, said, "More notes today?"

Ouch!

Now she didn't say this with a tone or with an attitude, but she was merely making an observation … one that hit me pretty hard.

THE MORAL

Starting your class in an engaging way is a key component to a successful class period. What I failed to understand that first year is that I had things backward. Rather than boring my students early with loads of science facts and figures and then hoping to reel them back in with some excitement such as an experiment or demonstration at the end, I now understand that the first 10 minutes of class often determines the success or failure of the entire class period. And if you can get your students to buy in to what your selling within those first few minutes, they will follow you wherever you take them for the rest of the period.

STEPS FOR SUCCESS

1. ADMIT SLIP, PLEASE!

As students leave your room, hand them a 3 × 5 card and let your students know that this is their "Admit Slip" for tomorrow's class. Explain to them that their slip should include one comment ("One thing I learned from my homework was …") and one question related to the homework ("One question I have about my homework is …"). In addition, let them know that they should be prepared to discuss whatever they write down.

The next day, stand at the door and collect the slips. As the students hand them to you, pick five comments that you think best summarize the main science concepts from the homework. When class starts, read these aloud as a way of solidifying the essential parts of the homework. You can also ask clarifying questions from the information they provide to check for understanding. When possible, allow the student who wrote it to lead the discussion.

Next, pick five questions from the slips to address aloud. Since there will often be many similar questions, this allows you to clarify and explain any hazy topics from the homework. Again, try to let the students play an active role in this by having them discuss the question with their classmates.

Please note that this assignment can be used both as a whole-class or small-group assignment depending on your students.

WHY THIS WORKS

First, it puts the kids in charge of the learning and the teaching. While we as teachers do our best to get inside the minds of our students, we sometimes miss the mark in seeing the work through their eyes. This activity solves the "guessing" we do.

Next, while the class is essentially covering everything I would have written on the board, it is being done in a discussion that is more engaging and less overwhelming. Regardless of how conversational we try to make our classes, when students see a board filled with notes, they tense up. This is a great way to have students lead the discussion on misconceptions, and it allows you the opportunity to assess their understanding by the way they respond and participate in the discussion.

Finally, my students love being recognized for their questions and comments. It gives them a "buy-in" to the learning, and it makes them *want* to be part of the class discussion. One of my favorite things to see is when kids start trying to one-up each other with the quality of their questions. I absolutely love when a student turns in an admit slip and announces to me, "My question is awesome!" This activity brings the curiosity and questioning back to the forefront of the science classroom, and it allows students to drive the inquiry learning.

2. INNER/OUTER CIRCLE

Prior to class, write 10 or so questions on the board related to the topic at hand. (I usually connect it to ideas from the homework, but this can also be used as an anticipatory activity to a new unit.) It's important to use questions that require a discussion or explanation, as opposed to a quick yes/no/single vocabulary word answer, For example:

- Can scientists predict earthquakes? Why or why not?

- Was the change in yesterday's experiment a physical or chemical change? What observations do you have to support your answer?

- Would you rather have mitosis that is out of control or meiosis that is out of control? What is the number one reason that made you choose your answer?

Next, have the kids stand up and move their desks to the outer portions of the class. After this, have half of the kids stand up and create a giant circle (facing outward) in the middle of the room. Ideally they should be about an arm-length away from each other, but this doesn't always work in larger classes. This is the inner circle. Finally, have the other half of the students stand up and face one member of the inner circle. This is the outer circle. This is diagrammed below in Figure 4.1.

Figure 4.1. Inner and Outer Circle Setup

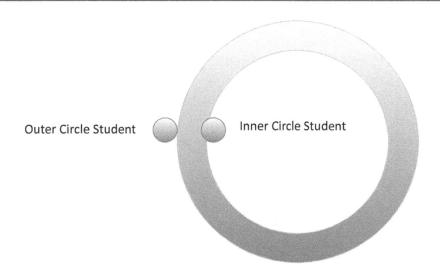

Once the inner and outer circles have been established, position yourself in the middle of the circle. Next, let the students know that the person in front of them is their discussion partner. Students will generally stand in front of one of their friends, so they will usually be psyched when you say this. However, if it is early in the year, I often ask that they introduce themselves to their partner. Once this is all in order, ask the first question from the board.

Give the students 30–60 seconds to talk with one another about the question. Be prepared—it's going to be loud. And while your teacher instincts may kick in at this point, resist the urge to tell them to be quiet. Interactive learning is supposed to be loud!

Once time is up, I have the students thank one another. I then ask the outer circle to rotate one spot to their right. Once there, they introduce themselves to their new partner, and you throw out question number 2. Repeat until all your questions have been discussed.

WHY THIS WORKS

Students love to talk, and this gives them that chance in a way that is meaningful and educational. Additionally, by putting the questions on the board *and* saying it out loud, both visual and auditory learners will have the benefit of receiving the question in the way that works best for them. Finally, students are hearing a different perspective on every question. This, to me, is the greatest strength of this activity. Rather than hearing the same ideas from two or three people over and over (as happens in small groups), students are continually hearing fresh voices and viewpoints. This gives students the opportunity to engage in scientific discussions where they can use data and observations to support their answers. This type of reasoning supports the discourse that is common in the science community and mirrors effective argumentation strategies that are essential to the science process.

From a teaching perspective, by listening to the responses given, you will discover what concepts are, and perhaps more importantly, are *not* being understood. This will then allow you to later emphasize any concepts that need additional clarification, and address misconceptions without directly pointing out a specific student in the group setting.

Again, let me stress that this one is going to create some noise. And while it might seem chaotic to someone who passes by your room as you are doing it, this activity not only gets the students involved in their learning, it also allows you assess the content knowledge of each student.

3. THE TOP 10

As class begins, divide your class into 10 small groups. Once the groups have been established, have each group spend two minutes creating a list of 10 words (single words, not phrases) that best summarize or relate to the topic at hand. Students should discuss and justify their choices with one another.

Next, have each group select the *one* word from their list that they feel is most important, and have them write it on the board. Let them know that there can be no repeated words on the board!

Finally, after all the groups have shared their word, go through the list, having each group explain the importance of their word. When they are finished, ask the other groups how many of them had that word as one of their top 10 words.

WHY THIS WORKS

First, it gets the kids talking about the subject. Plus, it challenges them to critically think about and justify why they selected the word(s) they did.

Next, it allows you the opportunity to assess what the students understand in a non-test manner. After each word has been discussed, you can also add your insights to the subject. And if there are any concepts left out, you can easily add your own words to the list to make sure the students get the information you need them to have.

Additionally, it allows students to see similarities and themes that exist within the science concepts that are addressed. For example, just because you had an experiment with DNA and followed it up with an activity on genetics, it doesn't mean the students see the connection between the two topics. By having them share important words, it can spawn a discussion on how those words are related to help students see the connections that exist. When looking for themes, crosscutting concepts listed in the *Next Generation Science Standards* provide an excellent resource for overarching themes in science that you can draw from to create connections.

I like to use this activity as part of an end-of-the-unit review. While it can be used for smaller segments of learning, make sure you have enough information to cover at least 10 key elements. It's better than your standard, "Here is a list of words to define" for the review, and really helps to emphasize the thematic connections and relevancy in science.

4. CIRCLE TIME

As a homework assignment, have the students write out three "would you rather" questions. Two of these questions should be related to your topic, while the other one is up to the student to make up.

The next day, as students walk into the door, collect these questions in a plastic bag. Once class has started, move the desks to the outer portion of the room, and have the entire class (and you) stand in a circle. Reach into the bag and read one of the questions. If students agree with option A of the "would you rather" question, have them step into the center of the circle. Those who opt for option B would remain where they are. Allow 20–30 seconds for kids to discuss why they feel the way they do, emphasizing the use of data and observations to support their reasoning.

Depending on how animated the kids are, you should be able to get through 20–25 questions in 10 minutes.

WHY THIS WORKS

Put simply, circle time works because it's fun. Students love justifying their answers, and they really have to make some complex comparisons in order to make their questions effective. Beyond academics, this activity allows students to see the commonalities they share with their classmates. Plus, students really get creative on the one they get to make up. Having a question related to your class followed by one related to pop culture, sports, or food keeps the kids excited to hear each question.

This activity really helps to start and maintain a discussion-based culture in the classroom, where students feel safe sharing their opinions, questions, insights, and ideas. It also helps to set the precedent and expectation that answers and opinions need to be supported by data and evidence. This type of argumentation is central to science. It also helps students recognize that science isn't just a set of facts, but instead is a community that is involved with interpretation and argument that evolves as more information becomes available.

Examples of circle time questions include:

- Would you rather be friends with someone who is described as stable or someone who is described as reactive?
- Would you rather live in an area that is prone to tornados or an area that is prone to earthquakes?
- Would you rather be a tree or a rock?
- Would you rather live forever or never feel pain?
- Would you rather be described by numbers or by words?
- Would you eat only protein or only sucrose?
- Would you rather continuously evolve or stay the exact same?

WHAT DOES SUCCESS LOOK LIKE?

Those three simple words, "More notes today?" completely changed the way I begin my class.

While I initially made sure to not take up *all* period droning on in front of my students, Kristen's simple comment made me realize that taking notes was my standard way of starting class each day. And yes, I would lead discussions, guide presentations, and oversee projects (among other things) with the rest of the period, but that one sentence made me realize that no matter how cool things turned out later in class, the first 10 minutes of my class were a complete and total bore.

And the positive effects of these activities extend beyond the first 10 minutes. Once students have "bought in" to a class, they are much more willing to put forth their best on any forthcoming assignments or projects.

By implementing these starter activities into my daily routine, my class now begins on a positive note, one where students look forward to being engaged in the learning. I have learned that having a schedule where the starter activities are posted for the week actually gets students to run to my classroom so that they don't miss out on the discussion and interaction with their peers. The culture has grown so much that former students from years past even stop by on Wednesdays because they know it's going to be circle time, and they want to participate. Additionally, beyond getting the students excited, these activities also lessen the amount of prep I have to put into setting up my class. The bonus of this is the fact that while I'm putting in *less*, my students are gaining *more*.

RESOURCES FOR MORE INFORMATION

Finton, N. 2011. *Science question of the day: 180 standards-based questions that engage students in quick review of key content—and get them ready for the tests.* San Francisco, CA: Jossey-Bass.

A few minutes a day is all it takes to get students ready for the science tests! Use this collection of short, thought-provoking questions to introduce or review key topics, such as animal adaptation, ecosystems, weather, the solar system, matter, and energy.

Yoder, E., and N. Yoder. 2008. *One-minute mysteries: 65 short mysteries you solve with science!* Washington, DC: Platypus Media.

Not an ordinary mystery book, *One-Minute Mysteries* makes science fun! Each mystery (solutions included) exercises critical-thinking skills while covering Earth, space, life, physical, chemical, and general science. A bonus section includes five mysteries from their upcoming title in the series, *One Minute Mysteries: Solve 'em With Math!* This entertaining and educational book is great for kids; grown-ups; schools; educators; homeschoolers; and anyone who loves good mysteries, good science, or both.

Wolke, R. L. 2000. *What Einstein told his barber: More scientific answers to every-day questions.* New York: Dell Publishing.

Do you often find yourself pondering life's little conundrums? Have you ever wondered why the ocean is blue? Or why birds don't get electrocuted when perching on high-voltage power lines? Robert L. Wolke, professor emeritus of chemistry at the University of Pittsburgh and acclaimed author of *What Einstein Didn't Know*, understands the need to ... well, understand. Now he provides more amusing explanations of such everyday phenomena as gravity (If you're in a falling elevator, will jumping at the last instant save your life?) and acoustics (Why does a whip make such a loud cracking noise?), along with amazing facts, belly-up-to-the-bar bets, and mind-blowing reality bites, all with his trademark wit and wisdom.

If you shoot a bullet into the air, can it kill somebody when it comes down? You can find out about all this and more in an astonishing compendium of the proverbial mind-boggling mysteries of the physical world we inhabit.

Arranged in a question-and-answer format and grouped by subject for browsing ease, *What Einstein Told His Barber* is for anyone who ever pondered such things as why colors fade in sunlight, what happens to the rubber from worn-out tires, what makes red-hot objects glow red, and other scientific curiosities. The book also includes a glossary of important scientific buzz words and a comprehensive index.

Mitchinson, J., and J. Lloyd. 2007. *The book of general ignorance: Everything you think you know is wrong.* New York: Crown Publishing Group.

Think Magellan was the first man to circumnavigate the globe, baseball was invented in America, Henry VIII had six wives, and Mount Everest is the tallest mountain? Wrong, wrong, wrong, and wrong again.

Misconceptions, misunderstandings, and flawed facts finally get the heave-ho in this humorous book of reeducation. Challenging what most of us assume to be verifiable truths in areas like history, literature, science, nature, and more!

The Book of General Ignorance is a witty "gotcha" compendium of how little we actually know about anything. It'll have you scratching your head wondering why we even bother to go to school.

Revealing the truth behind all the things we think we know but don't, this book leaves you dumbfounded about all the misinformation you've managed to collect during your life, and sets you up to win big should you ever be a contestant on *Jeopardy!* or *Who Wants to Be a Millionaire?*

Besides righting the record on common (but wrong) myths like Captain Cook discovering Australia or Alexander Graham Bell inventing the telephone, *The Book of General Ignorance* also gives us the skinny on silly slipups to trot out at dinner parties (Cinderella wore fur, not glass slippers, and chicken tikka masala was invented in Scotland, not India).

CLASSROOM MANAGEMENT 101

EASY STEPS TO BIG SUCCESS

Thuis chapter focuses on common classroom management techniques used by science teachers that allow for inquiry and hands-on learning in a controlled environment. The chapter looks at how young teachers can establish routines that allow them to apply the pedagogy highlighted in secondary science methods texts in a personalized and productive way.

THE STORY

The first week of school is always tough. Besides all the planning that goes into starting a new year, there is also the issue of getting to know all my new students (along with getting used to waking up at 6:00 a.m. again).

After a couple days of introductory activities (including "two truths and a lie" and "one secret no one knows about me is …"), I always like to wrap things up by asking the kids what they want to get out of the class.

Usually, most answers center around getting a good grade and, for the more adventurous students, leaning something along the way. But a few years back when I asked this question, Andrew's answer caught me by surprise.

"So what do you want to get out of this class?" I asked during the first week of school.

"I just want to not get in trouble by my mom, and I definitely don't want to end up in the principal's office," answered Andrew, a student who was new to the community and already demonstrating clear personality and opinions.

How awesome is that? There was no "I want to be the best student I possibly can be!" sucking-up happening from this kid. He just wanted to not get in trouble!

And because it was so awesome, I decided to roll with it.

"Sounds like we both have the same goals for this year. See, I don't want to get in trouble with your mom either, and I, too, *definitely* don't want to end up in the principal's office. I tell you what, Andrew … I'll help you avoid both those situations if you help me do the same. Deal?"

Over the next week, Andrew and I made a few verbal agreement to help guide us in our quest, including

1. He would turn in all his assignments, but I couldn't freak out if he didn't complete them all.

2. If he failed a test, I wouldn't e-mail his home. He would be responsible to tell his mom.

3. When he did well on something, I would e-mail his mom and praise his effort.

4. If he misbehaved in class, he would have the opportunity to explain his actions rather than immediately receiving a punishment.

5. At the end of every week, he would have to tell me one positive thing he did that week.

And do you what happened? Andrew and I had a great year.

Now I would lying to you if I told you I didn't have more than a couple meetings with Andrew and his mom, but I am proud to say that we both avoided the principal's office … at least for that year!

THE MORAL

Andrew was a kid who was used to getting into trouble. In fact, from what other teachers said, he almost fed into the behavior. But when I flipped the situation from him getting into trouble into **us** *not* getting into trouble, he really stepped things up.

I think Andrew, like many kids, just wants to know that someone is there to help them out. And once they know this, they are willing to do their best to live up to whatever expectation you have for them.

The point is this: If students know that you are dedicated to their success, they will do everything they can to not let you down. The following steps are suggestions that should re-emphasize and build on the lessons learned from your teacher preparation program to help you find the classroom management techniques that work best for you.

STEPS FOR SUCCESS

1. BE CLEAR ABOUT EXPECTATIONS.

One of the most difficult, yet most important, things to establish within the first few days of school is your class expectations. When students walk into a new class, one of the first things they try to figure out is what kind of teacher you are going to be. Are you going to be the nice, easy type or the mean, tough type? Are you going to be the happy, laugh-a-lot teacher or the never smile and grumble teacher? These are the issues that kids want to know! And it only makes sense, as we do the same thing as teachers. When you interviewed at your school, how many of you tried to draw conclusions about the environment of the school from your first interactions with the other teachers and administrators? It's a common approach to gauge your surroundings so that you can determine how to act in your new environment.

Try to not make their detective work difficult, and be upfront with your expectations by setting a tone and path to success that all students understand. This can include having the rules posted in your syllabus, on your website, or even posted on the wall. And discuss them with your students as well to make sure you are all on the same page. All of this will help you establish yourself in your new community and create a better classroom culture where students feel safe and know what to expect. To be clear, specific and direct instruction about expectations is really the best approach.

As mentioned above, you should set a high standard for *all* students, regardless of what their "file" may say. Too often, teachers have preconceived notions about students based on anything from how a sibling acted to what another teacher says. In essence, we sometimes come to a conclusion about a student *before we even know them* based on rumors.

Now think about that for a second … isn't that what we tell our students *not* to do?

Let them know where you set the bar, and how they can get there. It's the best way to start off the year, and it gives everyone a clear set of expectations that can be referred back to throughout the year if things start to go off course.

2. BE CONSISTENT WITH RULES AND PROCEDURES.

Kids notice things. And by things, I mean *everything*! If you haven't figured this out yet, give this a try: At the end of one of your classes, hand out a little piece of candy to each of your students as they walk out the door. Don't say anything … just give it to them on their way out.

As you do this, make sure that the kids from your next class see what is happening. When they do, one of two things will happen.

- They will ask you for candy as they walk in.
- They will ask you for candy when the bell rings.

Why will they do this? Because they expect the same thing to happen to them that they saw happening to their fellow classmates.

The same goes for the rules and procedures in your class.

Here's an example. If you say that turning in late assignments results in a 25% reduction, then you have to stick to that for *all* students. Yes, this even includes the students who will throw their "buts" at you such as

- "But I couldn't get it to print and it's only five minutes late!" 25% off.
- "But my I couldn't get my locker open!" 25% off.
- "But my dad's gonna kill me if I get a bad grade!" 25% off.

Like the candy scenario, if you do it for one, you have to do it for all. The flip side to this is if students know they can't get away with any of their "buts," the "buts" quickly disappear.

This concept applies not only for procedures within your classroom, but also to any department and school policies that are pre-existing. Remember, if you and the school believe in something enough to make it a rule, then you better believe in it enough to enforce it and be able to explain it beyond "That's just the rule." It's helpful to go through and have justifications for your expectations, and to share those openly so students understand. For example, I have a rule about not eating in the classroom. This is not because I am a mean and heartless teacher who doesn't want my students to grow so they can never be taller than me, but because we work with chemicals in the classroom space. Cross-contamination is a real concern, and one that could lead to students being out of my classroom for an extended period of time. I need each and every student present to be able to have the full community of perspectives and experiences, so I am not willing to put their health in jeopardy. This makes the "no eating" rule a lot easier to understand for kids, demonstrating that it comes from a position of concern as opposed to a position of power-hungry authority.

Added to this, it's important to recognize that you need to follow your rules as well. If it's a concern for the kids not to be eating in class, then you shouldn't be eating in classroom either. It creates a clear double standard, and really diminishes the severity of the threat and the importance of the rule. Think about it from your own perspective. If you have a principal who says teachers can't wear jeans because it looks unprofessional, but then wears jeans himself on a daily basis, do you think that's fair and justified? The golden rule holds true for students: Treat them as you would want to be treated.

From my experience, when you are consistent with your actions, both holding kids and yourself accountable, kids quickly learn what they can and can't get away with. And the craziest part is that they really don't mind, as long as it is consistent. Now that's not to say that they all *agree* with your rules, but rather that they at least *understand* them.

3. CONNECT WITH STUDENTS BEYOND THE CLASSROOM.

Kids will walk into your classroom and see you as a teacher. You need to help them realize that you're a person who teaches.

What's the difference? A teacher is only a teacher. A person who teaches is someone who has a life both in and *out* of the classroom.

I like to start class each Monday by spending five minutes talking about what happened this weekend. Topics I've covered include:

- A movie someone saw
- A concert someone attended
- A news event
- A popular song
- A YouTube video
- A funny commercial
- A TV show
- A sporting event
- A run I went on
- A favorite restaurant

Students are much more likely to connect with you as a person versus you as a teacher. And once this connection happens, they will work harder for you. For more ideas about how to connect with students outside of the classroom, see the chapter on Coaching and Community.

4. HAVE A SCHEDULE ON THE BOARD THAT CLARIFIES THE GOALS AND ACTIONS FOR THE DAY.

If you have ever attended a daylong meeting where they didn't provide a schedule, list of goals, or overarching ideas, it probably only took about 10 minutes before you started checking e-mail or started making your grocery list. It's fairly well understood that it's hard to be invested in a process such as learning without having a clear objective or where you are trying to get.

Using a schedule helps to provide this structure for students. Students are more likely to follow your leadership in a class when they know where they are going. The more specific you are in outlining the plans and expectations for the class, the more smoothly your class will run.

The schedule also helps create a path for students to follow if they finish early. Students are going to finish at different times based on their prior knowledge, efficiency, lab skills, and so on. Although it would be ideal for those students to engage with another group, sometimes that can be more of a distraction than a benefit. With a schedule on the board, students know that the next step is to start their conclusion (which is also their homework), and it gives them a clear task that helps them remain focused on the science learning that you designed.

An organized plan can also be helpful for creating a sense of urgency and commitment from the classroom community. There have been many times that I have had students stop a discussion from going on a tangent by reminding their peers (or myself) that we have three more tasks left to achieve on the schedule, and therefore, it would be best to get back on track. When students know the plan, they can help you and their classmates stay the course by eliminating classroom distractions.

Another suggestion I'd offer is to read the schedule aloud at the outset of class to establish the direction of the day for the visual and auditory learners in your room. You can also use it as a basis for the conclusion of your class by asking students a question like, "Looking back at the schedule, what did we learn/accomplish today?" This not only helps you check for their understanding of the information, but it serves as a great segue into any homework you are giving.

5. ALLOW STUDENTS TO ACT IN LEADERSHIP ROLES WITHIN THE CLASSROOM.

As a teacher, we often feel the need to be the center of all the learning in the classroom. This means we often speak first, are in charge of the exciting elements, and get to enforce all of the rules. Although this works for many students who have been groomed in a traditional K–12 classroom, you will encounter students each year who are not satisfied with being "foot soldiers." These are often the students who are known as troublemakers because they don't fit into the traditional follow-the-leader classroom experience.

These students can be turned from a liability to an asset by engaging them in leadership roles within the classroom. Examples of those roles include:

- Have students be in charge of the classroom supplies. This allows them to be active with both distribution and cleanup.
- Ask students to lead a demonstration. Go over the materials and practice with them ahead of time, then let them guide the demonstration.
- Have students be the journalist and record student observations on the board during the discussion. (This is a great tactic for teachers who have handwriting that is hard to read.)
- Designate students to be in charge of specific lab aspects, such as fire safety, to give them an additional focus.
- Engage students to act as leaders when you are out of the classroom. (This approach is detailed more thoroughly in Chapter 11, "Calling in Sick," p. 115.)

These roles work best when they tie into a student's specific talents or interests. For example, a student who is very active and has trouble sitting for extended periods is a great person to pass out handouts for you. Meanwhile, a student who is a recognized leader is an outstanding "lab manager" to give you the all clear when the lab is completely clean so the class can be dismissed.

If students are struggling with being contributing members of your class, ask them for their help by taking on these leadership roles. This allows the student to feel recognized for their talents, and it gets them to use those talents in a way that supports the learning in your science classroom.

6. USE A VARIETY OF CLASSROOM TEACHING APPROACHES.

No one wants to sit through a full class period of lecture, and that is why using variety in your teaching methods is so important.

It's helpful to have at least three different components in your daily class design. This could be a demonstration, a discussion, and a reading. It could also be a discrepant event, a writing reflection, and an experimental design task. By using multiple approaches to both pedagogy and learning styles, you can increase the excitement and engagement factor for your students. As engagement increases, classroom management issues decrease.

7. LET STUDENTS KNOW THAT THEY ARE PART OF THE SUCCESS OR FAILURE OF THE CLASSROOM.

This step related back to creating clear expectations for students, but it goes one step further by encouraging students to look at the class as *their* community rather than *your* science classroom. If you can engage them to create a classroom community, they will be the ones to support each other and keep each other involved in the learning. By openly recognizing that they play a vital role in the classroom experience, you give them both respect and responsibility. This will then translate into a stronger classroom culture.

WHAT DOES SUCCESS LOOK LIKE?

Successful classroom management starts and ends with how you run things. Being clear and consistent about expectations, using organized structure, and engaging students in leadership and community will help to create a positive classroom atmosphere.

To be clear, this can be one of the most difficult areas to master. It takes a great deal of time and experience with students to develop a clear understanding of what will be successful in your classroom. However, the best offense is a good defense, so using themes of clarity and student engagement can start you off on the right foot.

Like Andrew at the beginning of this chapter, students can sometimes put up a tough initial exterior. But I truly believe that all kids want to be successful, and it is your job to help them accomplish this.

RESOURCES FOR MORE INFORMATION

Wong, H. K., and R. T. Wong. 2004. *The first days of school: How to be an effective teacher, third edition.* Mountain View, CA: Harry K. Wong Publications.

The First Days of School offers anecdotal notes for new teachers and seasoned veterans on how to successfully facilitate your classroom in the first few days. These strategies will set the stage for the degree of success in your classroom over the course of the school year. This text includes suggestions on classroom management, lesson planning, and building a rapport with students.

O'Hanlon, T., ed. 2002. *Innovative techniques for large-group instruction.* Arlington, VA: NSTA Press.

Size *does* matter. When you're faced with a class of 50, 150, or even 250 college students, it's tough to head off boredom—much less promote higher-order thinking and inquiry skills. But it's not impossible, thanks to the professor-tested techniques in this collection of 14 articles from the *Journal of College Science Teaching*.

The book starts by examining what research shows about the effectiveness of popular teaching styles. (Surprise: Lectures don't stimulate active learning.) From there, the authors offer proven alternatives that range from small-scale innovations to completely revamped teaching methods. Suggested strategies include using quizzes in place of midterms and finals, student forums, interactive lectures, collaborative groups, group facilitators, and e-mail and computer technology.

The contributors write in first person, making the book as readable as it is practical. One of the most thought-provoking chapters is called "Are We Cultivating Couch Potatoes in Our College Science Lectures?" With this book's help, you can answer that question with a definitive *no*.

Emmer, E. T., and C. M. Evertson. 2012. *Classroom management for middle and high school teachers, 9th edition*. Upper Saddle River, NJ: Pearson.

Dealing with student misbehavior and encouraging student motivation are two of the most important concerns for new teachers. *Classroom Management for Middle and High School Teachers, 9th Edition*, provides new and experienced teachers with the skills, approaches, and strategies necessary to establish effective management systems in the secondary-school classroom.

Based on 30 years of research and experience in more than 500 classrooms, the newest edition of this best-selling text presents step-by-step guidelines for planning, implementing, and developing classroom management tasks to build a classroom environment that focuses on and encourages learning. Students can apply what they learn as they review and complete the examples, checklists, case study vignettes, and group activities presented in each chapter.

Powell, A. S. 2009. *The cornerstone: Classroom management that makes teaching more effective, efficient, and enjoyable, 2nd edition*. New York: Due Season Press.

Do you constantly repeat instructions? Are you fighting a never-ending paper battle? Strong classroom management is the cornerstone of effective teaching, and this book will help lay the foundation for everything you want to accomplish professionally:

- Designate a place for *every* type of classroom material
- Turn needy, disorganized children into self-reliant, responsible students
- Develop and teach *any* classroom procedure
- Train the class to follow along, stay on-task, and work together
- Use fun teaching techniques that help you assess student learning
- Eliminate homework hassles and parent miscommunications
- Prevent burnout by enjoying and growing with students
- Construct a self-running classroom that frees you to teach!

Using actual classroom photographs, forms, and dialogue examples, *The Cornerstone* will show you how to design instructional routines that facilitate learning. It will guide you through each step of communicating and reinforcing your expectations. Learn how to create a vision for your classroom and *teach*.

LAB SAFETY

MORE than JUST GOGGLES

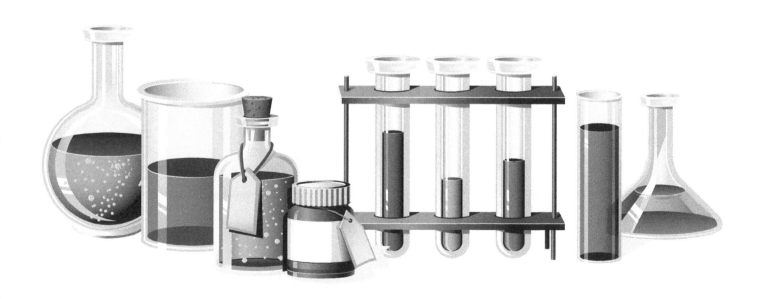

Every science classroom I have ever visited has always had a safety rules poster clearly displayed. As a new teacher, I assumed these were issued to every science teacher upon graduation. Hands-on science requires safe practices, a message that was resonated loud and clear in my methods course. This chapter focuses on how safety in a science classroom is more than a poster of rules, and how it can make the difference between an amazing classroom experience and tragedy.

THE STORY

With the fire alarm still going off, I find myself standing on the soccer field behind the school. I cringe as the fireman approaches me with my principal by his side.

"So can you tell me how the fire started?"

"I was making fireworks, sir." It was the end to a horrific day, and potentially my career.

My school was offering summer academic courses for the first time. This program was designed to allow students to engage in topics of interest in a less formal academic setting. Teachers were encouraged to create weeklong courses around topics beyond the traditional classroom. I had designed a class about "Toy Store Science." This course focused on the science behind common toys like silly putty, gliders, and card tricks.

With the July 4th holiday looming, I had also incorporated flame tests to look at the chemical properties of elements on the periodic table. In an effort to translate the experience into application (and at the prompting of my summer students), I had decided we would attempt to make a sparkler.

Finding recipes for homemade sparklers didn't take long. I even found some that were presented in a "lab report" format, making me feel all the more confident in my choice to support the students in their desire to create fireworks. After gathering together the required materials, I e-mailed the parents to let them know about the planned experience. I wanted to give them an opportunity to ask questions and give feedback before we embarked on this new experiment. In the e-mail, I explained that there would be no sparklers sent home from the course, and specific elements that required preparation ahead of time would be labeled with full chemical names to highlight the chemical properties of the experience. I even brought in a couple of sparklers from the local grocery store to do a compare and contrast with the class.

I prepared all of the materials and arrived early to complete the entire procedure myself prior to attempting it with students. I didn't experience any issues, and the final product ended up looking exactly like the sparklers that I had seen in the online video resource that I had found during my search. I felt confident that this experience was going to be amazing.

Thirty minutes after the sparkler experiment had started, I find myself racing to turn off gas jets and open windows as a beaker spews smoke and sparks into the lab. As I grab the fire extinguisher, a second beaker goes up in flames (from the sparks coming from the initial beaker combustion), and I start spraying a fire extinguisher for the first time in my life.

With the lab tables now covered in fire retardant, I quickly remove all other potential combustibles from the area. As my heart pounds I scan the room and survey the damage.

At this point, you might be asking yourself, "Wait … *Where are the students?*"

THE MORAL

Lab safety is of the utmost of importance for science teachers. As a profession that engages with hands-on experiences, we are responsible for the safety and care of the students who are in our classrooms. This is a heavy responsibility, and should be a primary consideration in the design and implementation of each lesson, activity, demonstration, or lab in your classroom.

Although I was fortunate that I came from a science background as an undergraduate and had experience with lab safety expectations from my college courses, many science teachers have not had this background when entering the classroom as a new teacher. While trying to learn about pedagogy, differentiation, assessment, and classroom management (among other things), lab safety can often be marginalized into a quick lecture or PowerPoint in your science methods class.

While time constraints often dictate a less-than-adequate approach to lab safety, science teachers are incredibly fortunate to have a wide range of safety resources available to them (including and beyond lab safety posters). New teachers can, and should, use these to support their knowledge of best safety practices.

STEPS FOR SUCCESS

Rather than describing the specific skills and content knowledge that you need for lab safety, this section will instead highlight local and national resources available to help you develop a successful lab safety program. Steps to take to improve your lab safety include the following:

1. TALK WITH OTHER TEACHERS IN YOUR SCIENCE COMMUNITY AND YOUR PRINCIPAL.

Although not always an exhaustive resource, your fellow teachers can help you to understand the school community expectations of lab safety. This can also be a chance to create cohesion about the expectations for students that can translate into a clear list of rules for students to know and follow. Questions to ask your colleagues and administrators include:

- Where is the safety equipment at our school?
- What procedures do we have in place if a lab safety incident occurs?
- How do I communicate and document an incident?
- Which adults are trained in first aid and CPR? Where can I get that training?
- Do we have a lab safety contract that is standard for all students in all classes?

2. BECOME TRAINED IN LAB SAFETY.

Many schools, districts, and states require lab safety training for all science teachers. If a course is not available in your area, Flinn Scientific (*www.flinnsci.org*) offers online courses that educate science teachers about safe practices and policies.

3. MAKE SAFETY RESOURCES AVAILABLE IN YOUR CLASSROOM.

There are many places where you can get lab safety resources, such as the American Chemical Association, science supply sources that sell lab equipment, and the National Science Teachers Association. These resources are for both you and your students. It's a great resource to have for student discovery when they wrap up early. Ask them to find lab safety violations and suggestions for improvement to keep them invested in the safety of your classroom.

4. KEEP A LIST OF THE COST OF THE LAB EQUIPMENT AND SUPPLIES IN PLAIN SIGHT.

During labs, you can refer students to the list, and have them report back to you about the cost of replacing the lab set up at their lab table. It's a little thing that students can forget when engaged in experiments. Labs cost money (see Chapter 3 about budgets), and helping them make that connection helps students take ownership of the lab.

5. CONTACT YOUR DISTRICT AND STATE OFFICES OF EDUCATION AND ASK FOR SPECIFIC GUIDANCE AND RESOURCES THAT ARE APPLICABLE TO TEACHING LAB SAFETY IN YOUR STATE.

District and state offices are great resources for knowing about specific laws and regulations that govern your classroom approaches, as well as what your legal responsibility is as a classroom teacher.

6. STAY CURRENT WITH LAB SAFETY.

Check your safety equipment regularly, and continue to take safety related courses throughout your teaching career. In the same way you stay current on best teaching practices for science, pay the same attention to student laboratory safety practices so you can offer your students a safe environment for exploration and inquiry.

WHAT DOES SUCCESS LOOK LIKE?

Back to the earlier question of, *Where were my students?*

They had followed lab safety procedures, and were lined up outside of the classroom (safety goggles still on their faces) as I attended to the sparking beakers. One designated student had gone down to tell the office, and another went down the hall to notify the other summer class about what had happened.

Even though the course was only for one week, I had started the experience by covering laboratory safety. We had spent time exploring all of the appropriate safety equipment, and had gone through different scenarios about what could happen in a lab setting. Before we began the experience in class, we had talked about the fire hazards and what steps we would take in case of emergency. (With summer students, my rules boil down to "Your responsibility is to let me know if there is an issue, and then you may calmly leave the classroom and line up outside the door." This helps to avoid any chaos or hero mentality amongst students who feel they are invincible.)

I had taught lab safety to every single student that I had encountered as a science teacher, thanks to guidance from an outstanding mentor teacher during my student teaching. For five years and countless students, I had *never* had a single issue beyond broken glassware. Going over lab safety procedures was a step I took because I knew it was right, but I had never had any reason to draw upon that knowledge for major action. As I looked from the fireman to the students seated safely and unharmed on the field, I was grateful that I had taken the time to instruct them on lab safety. Not a single child was injured in the incident, and the smoking beakers had been put out before any major damage had occurred.

Looking back, was making sparklers a good idea? I'm going to say no, and I now realize that I should have been much more thoughtful about the selection of my activities (including making sure the resources came from vetted sources). This story isn't to highlight my bad choice of activities, but to stress that lab safety measures were what kept my students safe. It is imperative that you take the time to educate yourself about lab safety so that you can run a safe and effective classroom that engages in hands-on activities ... without the fire department! It is your responsibility to be educated about these practices, so take a moment to follow the suggested steps to get trained through the resources that are outlined in this chapter.

RESOURCES FOR MORE INFORMATION

Roy, K. R. 2012. *The NSTA ready-reference guide to safer science, volume 3.* Arlington, VA: NSTA Press.

Safer science is a daily requirement for every teacher in every science classroom. Get up-to-date information from *The NSTA Ready-Reference Guide to Safer Science, Volume 3*. This volume is a collection of more than 40 quick-read "Safer Science" columns from *The Science Teacher*, NSTA's high school journal (plus some adaptable "Scope on Safety" columns from *Science Scope*, NSTA's middle school journal). As easy to read as it is practical, the book is chock-full of safety information, anecdotes, and advisories you can use every day.

The book covers a number of timely and important topics, such as

- systems to help prevent and control lab safety hazards, from eyewash showers to ventilation;
- standard operating procedures covering general safety precautions and safety in specific disciplines, such as biology, chemistry, Earth and space science, and physical science;
- personal protective equipment; and
- helpful safety-related NSTA position papers and internet resources.

Olliver, L. 2004. *Safety in the middle school science classroom* (flipchart). Arlington, VA: NSTA Press.

Put safety first by making *Safety in the Middle School Science Classroom* an essential part of every space where you teach science. Conveniently designed for hanging, this colorful flipchart ensures that you have, at a glance, the latest information for preventing safety problems in today's inquiry-intensive learning environment.

Focusing on the unique needs of grades 6–8, the flipchart provides the same features that made its elementary school equivalent an instant NSTA best seller. The front page has space for you to enter emergency phone numbers. A final checklist acts as a quick reference on some of the most important safety practices.

NSTA's highly regarded SciLinks notations reference up-to-date web pages on safe science. *Safety in the Middle School Science Classroom's* flipchart format combines the convenience of a calendar with the graphic appeal of a poster. Even your students will find its lavish and colorful illustrations irresistible. It's a quick-reference safety tool you'll flip through again and again.

Kwan, T., and J. Texley. 2002. *Inquiring safely: A guide for middle school teachers.* Arlington, VA: NSTA Press.

Not your average safety guide, *Inquiring Safely* is a uniquely readable resource from experienced teachers who know both middle school science content and how middle school students behave. The authors go beyond the standard rules and regulations to discuss safety concepts in the context of real classrooms—and to help you make students your partners-in-safety within an inquiry-based science curriculum.

New and veteran teachers alike can use *Inquiring Safely* to develop better approaches to equip labs, dispose of chemicals and other hazardous materials, maintain documentation, and organize field trips. Some chapters cover specific disciplines, such as physical science, chemistry, Earth science, and biology. Others deal with general topics such as supervising students' online activities, accommodating students with special needs, and working with volunteers. Special features include an unusually detailed index plus model student contracts and permission forms.

Like *Exploring Safely: A Guide for Elementary Teachers,* this essential book emphasizes a preventive approach to an up-to-date range of potential hazards. Given increased scrutiny of teaching practices and growing concerns about liability, *Inquiring Safely* belongs on the reference shelf of every middle school science teacher.

Texley, J., T. Kwan, and J. Summers. 2004. *Investigating safely: A guide for high school teachers.* Arlington, VA: NSTA Press.

Just as high school science is more complex than it is at lower grade levels, so are the safety issues you face in your classes and labs. Reduce the risks to people and places with *Investigating Safely*, the third and most advanced and detailed volume in NSTA's unique series of safety guidebooks for science teachers.

Some of the guide's 11 chapters deal with the special safety requirements of specific disciplines—physics, chemistry, Earth and space sciences, and biology. Others cover topics every high school teacher must grapple with, including equipping labs; storing and disposing of chemicals and other hazardous materials; maintaining documentation; and organizing field trips. You'll learn not only how to accommodate students with special needs but also how to make every student a partner in safer science.

Classroom veterans themselves, the authors have organized the book with practicality in mind. Safety concepts are discussed in the context of common situations in real classrooms. Sidebars and inserts in every chapter highlight and reinforce important material. Key information is selectively repeated in different chapters so you won't have to flip back and forth. And permission slips, student contracts, and other sample forms are included for adapting to your needs.

With scrutiny of teachers' practices and concerns about liability accelerating, *Investigating Safely* belongs on the bookshelf of every high school science teacher—and every science supervisor.

Flinn Scientific Lab Safety Courses (*http://labsafety.flinnsci.com/Home.aspx*)

Over the past 30 years, Flinn Scientific has successfully trained more than 100,000 high school and middle school science teachers in classroom and laboratory safety via safety seminars, safety workshops, and safety related e-mail messages.

Based on its years of experience helping teachers solve school science safety problems, Flinn Scientific has developed seven unique laboratory safety courses that you can view online absolutely free.

American Chemical Society (*http://portal.acs.org/portal/acs/corg/content*)

The American Chemical Society's website includes resources for students and educators regarding chemical health and safety.

Department of Health and Human Services, U.S. Consumer Product Safety Commission, Centers for Disease Control and Prevention, and National Institute for Occupational Safety and Health. 2001. *School chemistry laboratory safety guide.* Atlanta, GA: NIOSH.

The guide presents information about ordering, using, storing, and maintaining chemicals in the high school laboratory. The guide also provides information about chemical waste, safety and emergency equipment, assessing chemical hazards, common safety symbols and signs, and fundamental resources relating to chemical safety, such as Material Safety Data Sheets and Chemical Hygiene Plans, to help create a safe environment for learning. In addition, checklists are provided for both teachers and students that highlight important information for working in the laboratory and identify hazards and safe work procedures.

PARENTS

FRIENDS OR FOES?

This chapter looks at the role that parents play in your life as a new teacher. It will highlight back-to-school night, parent-teacher conferences, and how to engage parents as a resource for teaching science. Additionally, tips on how to establish boundaries and create open channels of communication will be addressed.

THE STORY

The first few years of my career, I was nervous any time I had to communicate with a parent. I *thought* I was doing a pretty good job in the classroom, but having complete confidence in your teaching ability is tough to do when you haven't had much time to test those abilities and get feedback from your community.

So every time I received a phone call or e-mail from a parent, it sent shivers down my spine. I mean, what if they asked me a question I didn't know the answer to? Or what if they were angry about a grade I gave their child and I was unable to fully explain why their child received it? Or what if … (fill in about a thousand other scary scenarios)?

Then one afternoon, shortly after school was let out for the day, a parent poked her head into my room. She was a parent that I had never met, and while her child was a decent student, she definitely wasn't one of my top performers. And as a relatively inexperienced teacher, the "pop-in" was something that I had convinced myself meant something bad was coming my way.

But then something weird happened: This parent told me that her child was really enjoying my class, and she just wanted to thank me for that.

Nothing more … nothing less.

THE MORAL

The entire conversation took less than 10 seconds, yet it got me thinking …

Maybe parents really aren't looking for the chance to jump down my throat. In fact, I began to wonder, how many of them are out there who are, do I dare say it, nice?

Teachers have a lot in common with parents. The most obvious similarity is that they both want what is best for the student. This commonality is often lost in the awkward and tense interactions that occur between the two groups in events such as back-to-school night and parent-teacher conferences.

It's important to keep in mind, however, that teachers can also intimidate parents. Your knowledge of science content, skills, and potential knowledge of household secrets often can make you seem difficult to approach. It is this lack of communication from *both* sides that makes the parent-teacher relationship strained and difficult to navigate, even for the most veteran teachers.

However, if you get back to the shared goal of supporting the students, it seems probable that working with parents should not only be possible, but something that can occur on a regular basis. Building on the simple pop-in interaction, I started to think of ways that I might be able to include and inform parents about what was happening in my class. Rather than fearing a confrontation *from* a parent, I instead turned my attention on ways in which I could collaborate *with* them.

STEPS FOR SUCCESS

1. BACK-TO-SCHOOL NIGHT

This is often your first opportunity to interact with parents, which can often come across similar to a first date. Both parties are trying to put their best foot forward, while also trying to learn about the other's intentions. This can be an incredibly intimidating moment, as most new teachers have not had an experience hosting a back-to-school night prior to their first year. Whether it is a formal all-school event or a casual meet-and-greet open house, consider these suggestions when planning your first "big" parent meeting:

- Take a moment to introduce yourself and share your credentials. Especially for a new teacher in the community, it's helpful to give your background to educate parents about your previous experience and training.

- Tell parents why you are excited to support their students learning science. Your passion for science education can be your greatest asset in setting a positive tone for the school year.

- Share student work or photos of the students. Many parents like to see evidence that their child actually participates in your class (this is even true for parents of high school juniors and seniors). Whether it's science safety posters the students created for the room, or a list of goals that your students created for the class community with their signatures, it helps to personalize the experience.

- Be clear on setting boundaries with your preferred method of contact. I was very surprised to receive a phone call at 11:30 p.m. asking for clarification on the instructions for a homework assignment during my first year of teaching. Even more surprising was that it was the first of many that occurred during that school year, let alone getting text messages asking for support. The next year, I clarified that the best way to reach me was by e-mail, and that my goal was to respond within 48 hours. It made it clear for the parents that this was the best way to reach me and helped me to be accountable for timely communication.

- Outline steps students can take to be successful in your classroom. Just like the students, the parents are curious about how to best support their kids. It's a great moment to point out the responsibilities and expectations you have for students, while giving parents clear instruction on how they can support that success at home.

- If parents want to talk about a specific student concern, it's best to ask them to contact you after the evening to set up a time to chat where you can focus on their questions. This will keep you free from being monopolized, but still allows you to give them a venue to continue the conversation.

- End with your goals for the school year. It's a positive culmination and sets a tone of achievement and growth for your classroom.

2. AN E-MAIL OR PHONE CALL A DAY KEEPS THE ANGRY PARENTS AWAY.

For the first several years of my teaching career, an e-mail or phone call from me to a parent meant bad news. A bombed test, inappropriate behavior in class, or not turning in homework was cause for the standard "I just wanted to let you know that …" e-mail or call.

The problem was, I rarely heard back from these parents. And if I did, it was generally a quick "Thanks for the heads-up. I'll talk to him/her about it." While I was trying to keep the parents in the loop about things, it often felt like my communication just created a larger divide between us.

A few years ago, however, I got the crazy idea to send home e-mails or make phone calls regarding the *positive* experiences I saw in class. A strong test, an insightful comment in class, or a great presentation were now the cause of the new "I just wanted to let you know that …" e-mail or call.

And the craziest part of this new concept? Over the years, I have heard back from the vast majority of these parents, including several who told me that this was the first positive e-mail or call they had ever received from a teacher. In one specific case, a parent of a straight-A student said it was the first student-specific contact she had *ever* received from a member of the school. Keep in mind that her child had been at the school for *several* years. She was even considered one of the school's top students, and yet her parent had never received feedback beyond the traditional parent-teacher conferences!

Sending e-mails or calling parents is something that every good teacher does, but it doesn't have to be a negative experience. By starting the school year off with a positive source of communication, it opens up a friendly relationship between you and them. Additionally, it lets them know that you are there to help their child succeed, not just to point out the flaws in them. And even when those bad e-mails/calls are needed (and they will be), you have laid a positive foundation for communication that shows you care for their child. This highlights that you both have the same goal for supporting the student, and it helps to put you and the parent on the same collaborative team.

3. BEGINNING-OF-THE-YEAR PARENT LETTER

One way to get the school year off to a positive start with parents is to give them the chance to brag about their child via a beginning-of-the-year letter.

It works like this: During the first couple weeks of the school year, I send home a "homework" assignment for parents to complete. (This can also be accomplished through an e-mail or at back-to-school night.) In this one-page letter, I encourage parents to tell me a little about their child regarding both academic and personal stories related to science. Additionally, I encourage them to include positive traits, accomplishments, and any funny stories they wish to share.

I also let them know that I will be reading these aloud in front of the class, so if they want to embarrass their kid a bit, this is the place to do it! As odd as this sounds, this is what gets many of the parents to "bite" on this assignment.

The last request I have is for them to leave the letter completely anonymous and genderless. Some parents accomplish this by simply saying "My child" or "This student," while others offer up the names "It" and "Child X".

Once I read the letter, everyone in the class tries to guess which student I am talking about. After he or she is identified, I usually ask a couple follow-up questions related to the letter to give them the spotlight for another couple seconds.

You will be amazed at how into this the students get! As you read, you will see kids pointing at each other or looking around the room for the kid with the reddest face. And while it does take some time (about two minutes per letter), if you spread it out over a month (reading two or three per day), it becomes much more manageable. Plus, it gives students something to look forward to each time they enter your classroom.

This assignment builds a successful relationship with parents in that it shows you want their help in getting to know their child better. As we all know, parents love to talk about their kids, and this lets them do so in a positive and engaging way.

Oh, and sometimes you will have parents who fail to submit a letter. I will often e-mail a reminder to those who forget (which happens quite frequently). If I still don't get anything, I give them one last reminder (usually at parent-teacher conferences). If all that fails, I give the student the chance to write a letter about themselves that shares some of their unique talents.

4. INVITE! INVITE! INVITE! (YOUR WAY)

Another way to build a positive rapport with parents is to invite them to participate in activities related to those within your classroom.

Now make no mistake; I am not suggesting that you start inviting parents into your classroom on a daily basis. In fact, I actually recommend just the opposite. After all, there is only a certain segment of your population that can visit during school hours, and as nice as those parents are, seeing the same three or four parents over and over gets to be a bit awkward, especially when they get comfortable enough to make themselves at home in your classroom. Plus, it can make the students feel awkward having a parent in their learning environment if you haven't checked with them first.

That is why I suggest inviting parents to community events related to your class that you oversee, not lead. For example, during an astronomy unit, find out if there are any local stargazing groups in your area who might guide a night under the stars for you and your students. Or maybe you are doing a unit on plants. Is there a local community garden looking for a little help for an hour?

Whatever the event, let parents know that you will be there, but that you will be there as a *participant*, not the teacher/leader. And while many of the parents will still want to talk to you about school, it will also create a learning community where you, students, and parents are all learning together.

When working with adolescents, one of the biggest challenges with engaging parents in activities outside the classroom can be the students themselves. They are in a time of life where they are establishing their independence, which can make it challenging to have them openly invite parents to attend these events. One approach to alleviate this issue is to issue a participant challenge. Let the students know if they can get a total of 50 or more people to attend the event, it will result in an added bonus of hot chocolate (or whatever your reward is) at the next event. By doing this, students will be quick to introduce you to their guests so they are "counted" toward the goal. At the same time, this gives them a sense of pride in getting more people to attend. Finally, it can also be a great instigator for increasing parent participation simply because parents are more likely to engage when they see that other parents are present.

Inviting parents to community events helps to foster a lifelong learning community within your school. It shows parents that you want them to be involved in what is happening in class, but it does so in a way that doesn't interfere with the actual teaching that is happening in your classroom. Parents need to know that you care about their children, and offering up activities for them to engage in will help solidify this.

5. OFFER TO TEAM WITH PARENTS.

Eventually, you will have to have a meeting or a phone call with a parent to address an issue that is not so pleasant. Whether talking about academics or behavior, these conversations can quickly become combative to the point where both you and the parents are trying to establish who is to "blame" for the student's struggles.

Whenever these conversations take place, there are some key tactics that can help alleviate the challenges:

- Engage the student in these conversations. By the time they have reached secondary school, students are being asked to be proactive and self-advocate in their learning environments. To support this goal, it's important that they are the central focus and voice in the meeting.

- Start the conversation by clarifying the goals of the meeting. Usually there is a specific topic to address (completing work, respecting peers, and so on), but make sure to identify the desired outcome and how you will help the student find success. This can be something as simple as saying, "My goal for this meeting is to establish some simple steps that will help your child be more successful. And while he plays a major role in this success, it's also important to note that I am here to help him achieve this." This simple articulation will often get you and the parents on the same side from the onset of the meeting.

- Take ownership of specific elements that you can support in the classroom, and ask parents how they see this translating to their expectations at home. For example, offer up a strategy you can use in your classroom, then ask the parents if this is similar to something they might already do at home. This allows them to see what you suggest as an educator, and then provide their own ideas for what would work in their household that is similar.

- Follow up with parents and students after the meeting. Are the strategies you agreed on working? Does the student feel better about their work and role in the science classroom? These follow-up pieces are helpful for gaining feedback from home, while also demonstrating your investment in your students. It also keeps the communication channels open for additional questions and concerns.

6. PARENT-TEACHER CONFERENCES

This is another venue where many new teachers don't have direct experience before being asked to host their first set of conferences with parents. Each school is going to have a different protocol, so it's best to ask your colleagues and mentor about how they approach this event to get a sense for what is expected of you as a teacher. After you have the protocol, a few helpful tips:

- Before you sit down for conferences, it's helpful to have student data and/ or observations to directly refer back to during the conference day. You may feel you know every student in your classroom, but if you have spent your day talking to 50 different parents you can find the details get fuzzy. Whether it's grade reports, a specific assessment from your class, or a list of strengths and challenges that each student generated to share with parents, it can be a helpful reference to have available for the conference.

- Another good step can be to touch base with another teacher who has the same students before meeting the parents. This gives you an opportunity to compare observations and look for similarities. It's a good way to confirm that you have accurate observations, and can be a good heads-up if you are sharing information that is inconsistent with other courses.

- During the conference, ask the parents what they have heard from their students about science class. If you teach adolescents, the answer may be "nothing," but it gives you a good avenue to start a dialogue about what has been happening in your class and how that can translate to their home environment.

- Try to have at least one "strength" and one "item for improvement." This can be a difficult challenge, most often for the students who are excelling in your classroom. It's a good reminder that our goal is to challenge *all* students, and articulating that challenge helps support differentiation in your instruction.

- Collaborate with parents to talk about goals for science. This is a great opportunity to talk about approaches that parents can take to support their student, while giving you an opportunity to gain insight about specific strategies that may help you with their student.

- If you have an angry parent (everyone does at some point), it's best to let them share their frustration and engage in active listening. Repeating their concerns such as "I hear you saying you are upset with their citizenship grade," helps them see that you are hearing their concerns. It can also help to diffuse the situation, and move back to a place of collaboration by following with "How can we work together to help your student improve?"

- If you have extra time at the end, you can always take time to give a heads-up on future topics and projects in your science class.

7. USE FACE-TO-FACE ENGAGEMENT WHENEVER POSSIBLE.

Sending an e-mail can be very easy and convenient, but it is not always the most effective avenue for sharing important information. Specifically, if the information deals with personal or serious concerns about their child, e-mails can often come across as cold and lacking in emotional investment.

Additionally, many people say things in an e-mail message that they would never say in person. I know this, because I sent a very angry e-mail to a restaurant once after getting food poisoning. When the manager called me back (since I forgot my phone number was in my automatic e-mail signature), I apologized for the hasty and curt feedback and welcomed his follow-up.

If you need to contact a parent for a serious issue, it's best to invite them in or call them about your concerns. Also, if you receive an angry e-mail from a parent, it's best to follow up by scheduling a face-to-face meeting. Meeting in person can diffuse the situation, and it can help you and the parent get back on the same page of supporting the student.

WHAT DOES SUCCESS LOOK LIKE?

Too often, teachers view parents as an obstacle rather than as an opportunity. I suggest that rather than waiting for a confrontation, be proactive by seeking out ways for *collaboration*.

More than anything else, parents want to know that you care about their child. By letting them be involved in this process, you will build a strong foundation that invites parents to be part of the experience.

Starting things off on a positive note with parents (see Figure 7.1) will lead to a productive and supportive environment for both the parents and you. Parents will appreciate your efforts to help their child succeed, and they will be more receptive to your suggestions if and when those arise.

Figure 7.1. E-Mail Templates

POSITIVE PARENT E-MAIL TEMPLATE

Hello,

I just wanted to drop you a quick note to let you know how impressed I was with _____ 's test/paper/project/participation today. She really worked hard to do well on this, and you (and she!) should be very proud of her effort.

If you get a chance, give her an extra pat on the back tonight.

Thanks,

The Science Teacher

BEGINNING-OF-THE-YEAR PARENT LETTER INSTRUCTIONS

Hello Parents,

As a way to get to know your child better, I need your help. Here's the scoop…

Please type a short (one page or less, double-spaced) letter describing your child. *Please do not give the student's name in the letter!* Detail his/her hobbies, talents, accomplishments, etc. I will read these letters out loud to the class, and they will try to figure out who it is based on your description. And please do not tell your child what you're writing! It's better when he/she gets to hear it for the first time in public.

This is a great chance to brag and maybe share some little known facts about your child. Please sign the bottom, then seal the letter with the student's name and period on the outside. You can also e-mail it to me to maintain confidentiality. *I will start reading the letters on _____,* so if you could get me the letters by then, that would be cool. And if you're willing to play along, it gets your kid 10 points. If you cannot participate, it won't hurt his/her grade.

Thanks,

Your Science Teacher

RESOURCES FOR MORE INFORMATION

Tingley, S. 2006. *How to handle difficult parents: A teacher's survival guide.* Waco, TX: Prufrock Press.

This book offers practical advice for teachers, presented with a sense of humor. The stress of dealing with difficult parents remains one of the top reasons teachers cite for leaving the ranks, according to the Center for the Study of Teaching and Policy. *How to Handle Difficult Parents* helps teachers learn how to cope more effectively.

Learn how to handle parents like these:

- Helicopter Mom, who hovers constantly, ready to whisk away any problem or inconvenience that might befall her child;

- The Intimidator, who wants what he wants and wants it now; and

- Pinocchio's Mom, who believes that her child, unlike every other child in the universe, never ever tells a lie of any kind.

You will also find out more about the Caped Crusader, Ms. "Quit Picking on My Kid," the Stealth Zapper, the Uncivil Libertarian, No Show's Dad, and the Competitor.

Whitaker, T., and D. Fiore. 2001. *Dealing with difficult parents and with parents in difficult situations.* Larchmont, NY: Eye on Education.

This book helps teachers, principals, superintendents, and all educators develop a repertoire of tools and skills for comfortable and effective interaction with parents. It shows you how to deal with the parent who is bossy, volatile, argumentative, aggressive, or maybe the worst—apathetic. It provides specific phrases to use with parents to help you avoid using "trigger" words, which unintentionally make matters worse. It will show you how to deliver bad news to good parents, how to build positive credibility to all types of parents, and how to foster the kind of parent involvement that leads to student success.

Henderson, A. T. 2007. *Beyond the bake sale: The essential guide to family/school partnerships.* New York: The New Press.

Countless studies demonstrate that students with parents actively involved in their education at home and school are more likely to earn higher grades and test scores, enroll in higher-level programs, graduate from high school, and go on to postsecondary education. *Beyond the Bake Sale* shows how to form these essential partnerships and how to make them work.

First published by the National Committee for Citizens in Education in 1986, *Beyond the Bake Sale* went on to sell more than 50,000 copies in nine editions. Packed with tips from principals and teachers, checklists, and an invaluable resource section, this updated and substantially expanded edition reveals how to build strong, collaborative relationships and offers practical advice for improving interactions between parents and teachers, from ensuring that PTA groups are constructive and inclusive to navigating the complex issues surrounding diversity in the classroom.

Written with candor, clarity, and humor, *Beyond the Bake Sale* is essential reading for teachers, parents on the front lines in public schools, and administrators and policy makers at all levels.

The book includes answers to these questions:

- What is a family-school partnership supposed to look like?
- How can schools and families build trust instead of blaming each other?
- How can involving parents help raise students' test scores?
- How can teachers relate to families who don't share their culture and values?

THE **GRADING DILEMMA**

BALANCING GREAT IDEAS WITH A MANAGEABLE PAPER LOAD

T his chapter centers on finding a balance between all the great lessons created in the classroom and all the paperwork said lesson leaves behind. Specific assessment approaches that can be easily implemented into any unit will be addressed, along with how to use digital documents to lighten the "lab book" load.

THE STORY

It's January, and I am sitting behind a mound of 12-page science exams that use multiple methods of assessment, including short answer, diagrams, word problems, and even a matching section. Although the students had just completed the tests the previous day, the nonstop barrage of "Have you graded our tests yet?" was in full effect.

"Oh, yes. I just snapped my fingers and they were finished!" My *Mary Poppins* reference is lost on them, and I am reaching a point of despair as I recognize that I am looking at a dreary President's Day weekend of nothing but grading exams. My excellent education in multiple forms of evaluation is coming back to haunt me in the form of hours upon hours of grading authentic assessment.

Later that afternoon, I haul the tests into the study period I help oversee. I look over at the other teachers and I see they are engaged in a discussion about the school's varsity basketball program.

How are they doing this? Don't they have mountains of papers, exams, homework, and projects to grade, too? Am I the only one who is assigning work this month? Is this just something that impacts science teachers and not the other disciplines?

As the bell rings, I attempt to gather the mountain of science exams to haul from the library to my classroom. *At least I will be working toward my goal of toned arms,* I think as I strain to get all of the papers off the table. As I reach over to get my computer bag, the load of papers suddenly lightens.

The English teacher, Mike, grabs half of the exams. "I thought science teachers were against killing trees," he says. "But looking at all of this, you seem to have taken an opposing viewpoint. Here, let me help you." I am grateful for the assistance, and we slowly make our way back to my classroom through the wave of students.

"Do you mind me asking what all of these are?" the English teacher asks as we navigate through the post-bell chaos.

"Physics exams. Don't you see the title, 'Phun with Physics'? It was their second cumulative exam. We are working on linking ideas across topics to see themes, so this covered quite a bit of material."

"So it seems," he says with a smile. "Are you grading all of this yourself?"

I knew it! They all had grading robots and were keeping mine from me until I had proved my worth as a teacher. "Yes, that's the plan, unless you are volunteering your Physics expertise?"

Mike pauses, and then asks a question that is about to shatter my teaching world. "Have you ever thought about the kids grading these?"

THE MORAL

As a science teacher, you spend a lot of time with your students trying to gain a better understanding of what they know, what they are learning, and what questions they still have. Assessment is a thematic strand that is addressed often and in-depth in many teacher preparation programs. Lessons and resources in formative and summative assessments translate from the college classroom to the K–12 classroom. There are entire books dedicated to assessment strategies for science educators, and because of this, this chapter is not going to go into all of the specific types of assessment and evaluation that can be used. For more information there are outstanding resources in your science methods texts as well as from NSTA Press about effective assessment in science. However, most of these resources focus specifically on the design and implementation of assessment in science and fail to address the time commitment that can be required to complete the evaluation.

As a new teacher, time is a very valuable resource. You are constantly planning new lessons; searching for valuable resources; organizing materials; engaging with students, parents, and colleagues; and occasionally eating and sleeping. If you employ authentic assessment and have moderate to large numbers of students, you can quickly find yourself drowning in a sea of lab books, practice problems, and quizzes. This paper load can be one of the largest drains on your time, and is often the one that has the largest impact on your time outside of school. New teachers are notorious for taking home hours upon hours of grading, and often look at the veteran teachers with confusion and envy. So how do you avoid the mountain of papers as a new teacher?

STEPS FOR SUCCESS

1. USE A GRADING RUBRIC.

Rubrics are incredibly helpful for both students and teachers. They help the students have a clear understanding of the assignment criteria, and they give you a format for providing a grade in a structured format. Detailed rubrics allow you to circle the statement that best describes the student work, giving feedback without having to write the same criteria statement on each paper. You can also ask students to grade their work using the provided rubric. They then will turn it in to you to see if they can accurately evaluate their final products and demonstrate their skills in self-evaluation.

2. DON'T GIVE BUSY WORK.

If the work is valuable only for the sake of keeping the students busy, it will be equally effective in keeping you busy to grade it. Be thoughtful about the assessment you are using with your students and what outcomes you expect to get from that assessment. Assessment should be beneficial to both you *and* the student, and work for the sake of having work does not accomplish that for either party.

3. STAGGER WHEN YOUR UNITS END AMONG YOUR VARIOUS PREPS.

The only thing worse than having a stack of tests to grade for one class is to have five stacks from three different sections! While it might seem like a good idea, don't start and finish all your units at the same time. This will just cause a major grading backup when each of them comes to an end (not to mention a major stress backup!)

4. USE TECHNOLOGY.

There are *many* new tools available through technological means for gaining instant data on student understanding. Whether it's using online Google forms to do polls, Survey Monkey to get specific data from students, or clicker systems to get instant feedback in class, technology can be helpful in generating quick evaluation tools that give you a general sense of the level of understanding in your class. Although these don't always provide data down to the specific student level, they can give you the general strokes of what is working and not working in your classroom instruction. As a result, you can make changes on the spot to address student needs.

5. USE DIGITAL DOCUMENTS.

There are many opportunities to allow students to engage in digital documents. This allows students to create collaborative work, such as lab reports with collective data, without the hassle of trading paper documents. Digital documents are also incredibly useful for reducing the amount of papers you carry, as well as for giving you a digital means to give comments and feedback. Options include Google Docs, Zoho, or ThinkFree.

6. HAVE STUDENTS EVALUATE ONE ANOTHER'S WORK DURING THE REVISION PROCESS.

One example of this is asking students to create a draft of their conclusion statement from an experiment. When revising, ask each student to document one thing their peer did well, and identify one question that they need to clarify. By getting feedback from their classmates, it helps give your students multiple perspectives for improvement. It also emphasizes the importance of community in your classroom, as well as helping the students achieve a better final product. Finally, it reduces the amount of time it would take you to complete the entire draft and final assessment on your own.

7. HAVE STUDENTS EVALUATE THEIR OWN WORK (INCLUDING TESTS).

Yes, I just said to let kids grade their own tests! This can be successfully completed in a discussion-based setting, where students have the opportunity share questions, concerns, and feedback with both you and their classmates. As a teacher, you should provide a general idea of what a "good" answer would look like. After modeling this, open it up to students who want to share their answers with the class. After hearing the answer, you lead a discussion about why that answer is correct or not. This not only creates a discussion about the answer, it also allows students to ask follow-up questions, as well as giving you an opportunity to address specific misconceptions that may exist among your students.

8. FOCUS ON THE PROCESS, NOT THE PRODUCT.

This is especially true with regard to projects and lab notebooks. Due to their size and weight, assignments like these are difficult to take home. As a result, most of these will be graded during prep periods. To cut back on grading time, create a rubric (see Step 1 for overview) that centers on identifying "big picture" concepts. Since these assignments have a much more limited time to be graded (due to not going home with you), pay attention to *how* they answer versus *what* they answer. This will help you get a better grasp of any concepts that might need to be explained in greater depth.

WHAT DOES SUCCESS LOOK LIKE?

"You let students grade their own papers? How is that even possible? Don't they change answers? What do you do with short answer? Don't the students go ballistic, since grading is the teacher's job?" My questions flowed out as my mind raced.

I had never considered having students grade their work, since I was never allowed to grade my own work during my K–12 experience. My questions came across as challenging and defensive, but they actually were more from a place of curiosity. I quickly backed away from the "20 questions" approach, and instead asked if I could come and observe Mike's classroom to explore this possibility.

I was amazed to see the structure, efficiency, and *learning* that took place in the English classroom as they assessed their own papers. Mike was skillful in setting up a structured environment where the grading expectations were clear. He outlined the specific elements and points associated for the short-answer questions, and students highlighted and marked up their responses to support the rationale for their final total score. He had created a safe culture for sharing, and students openly offered responses to be reviewed by their peers. They asked clarification questions about why their answer was incorrect, and many gained additional understanding by listening to their peers justify and explain their reasoning and evaluation. Rather than looking just at the grade, the assessment became the basis for discussion, and students were invested in continued learning as a way to improve understanding of both skills and content. Students were able to ask for a second review by the English teacher if they had additional questions, which he would follow up on outside of the group discussion. Mike still evaluated the students by reviewing their self-graded assessments, and he took the time to talk with those anyone who struggled with content or the grading process.

After seeing this process, was I ready to hand the Physics exams back to my students to grade themselves? Not quite. But I was ready to do a self-assessment with a shorter lab conclusion to get my class adjusted to the new system. I drew upon their experiences from their English course, and I created a similar structure to create cohesion across their classrooms.

Working in collaboration with Mike and my other colleagues, I was able to reduce the time that I was spending on grading while still gaining the data necessary to assess my instruction and provide specific student support. The reduced time grading translated into more time for planning and designing lessons, working with students individually, and engaging with my colleagues and community. It also gave me more time outside of school to have a better balance between my career and personal life.

As a new teacher, it's important to find as many time-saving activities as possible. Whether it's grading tests, planning lessons, or purchasing supplies (to name just a few), time becomes your most valuable resource. Managing it well can help to define your success both with students and in sustaining your passion for being a science educator.

RESOURCES FOR MORE INFORMATION

Atkin, J. M., and J. E. Coffey, eds. 2003. *Everyday assessment in the science classroom.* Arlington, VA: NSTA Press.

Make ongoing, classroom-based assessment second nature to your students and you. *Everyday Assessment in the Science Classroom* is a thought-provoking collection of 10 essays on the theories behind the latest assessment techniques. The authors offer in-depth "how to" suggestions on conducting assessments as a matter of routine—especially in light of high-stakes standards-based exams, using assessment to improve instruction, and involving students in the assessment process.

The second in NSTA's Science Educator's essay collections, *Everyday Assessment* is designed to build confidence and enhance every teacher's ability to embed assessment into daily class work. The book's insights will help make assessment a dynamic classroom process of fine-tuning how and what you teach… drawing students into discussions about learning, establishing criteria, doing self-assessment, and setting goals for what they will learn.

Keeley, P., F. Eberle, and L. Farrin. 2005. *Uncovering student ideas in science, volume 1: 25 formative assessment probes.* Arlington, VA: NSTA Press.

Before your students can discover accurate science, you need to uncover the preconceptions they already have. This book helps pinpoint what your students know (or think they know) so you can monitor their learning and adjust your teaching accordingly. Loaded with classroom-friendly features you can use immediately, the book is comprised of 25 "probes"—brief, easily administered activities designed to determine your students' thinking on 44 core science topics (grouped by light, sound, matter, gravity, heat and temperature, life science, and Earth and space science). The probes are invaluable formative assessment tools to use before you begin teaching a topic or unit. The detailed teacher materials that accompany each probe review science content; give connections to *National Science Education Standards* and *Benchmarks*; present developmental considerations; summarize relevant research on learning; and suggest instructional approaches for elementary, middle, and high school students. Other books may discuss students' general misconceptions about scientific ideas. Only this one provides probes—single, reproducible sheets— you can use to determine students' thinking about, for example, photosynthesis, Moon phases, conservation of matter, reflection, chemical change, and cells. Each probe has been field-tested with hundreds of students across multiple grade levels, so they're proven effective for helping your students reexamine and further develop their understanding of science concepts.

Lantz Jr., H. B. 2004. *Rubrics for assessing student achievement in science grades K–12.* Thousand Oaks, CA: Corwin.

The need has emerged among science educators for assessment and evaluation tools that will complement and extend traditional selected-response test items. The assessment tools within this packet were designed to address this need.

GETTING TO KNOW YOUR STUDENTS OUTSIDE OF THE CLASSROOM

COACHING AND COMMUNITY

This chapter focuses on interacting with students outside of the classroom to improve the effectiveness of your teaching. It will highlight engagements through coaching (athletics and academic groups like FIRST LEGO League and Science Olympiad), as well as community activities (including organizing a stargazing night, rock climbing, and movie night). The chapter addresses how to engage and organize these activities, as well as how to use them to your benefit in the classroom.

THE STORY

It's the start of an elective course in the new trimester. I have a group of science enthusiasts anxiously talking about their plans to win the upcoming state Science Olympiad competition. As they reminisce about last year's close call with the bottle rocket that *almost* won (meaning it took sixth place), they immediately begin talking about changes they must make in order to win the title this year. It is truly a new science teacher's dream class. All the students have passions to pursue STEM careers. They live, eat, and breathe science.

And then there is Adam.

A minute before the bell rings to start class, Adam sinks into the desk closest to the door (for a quick escape). It's clear from the two empty rows of desks between him and the other students engaged in plotting their future victories that this was not Adam's first elective choice. Adam has spent the first third of the year in his regular physical science class simply trying to get by without getting noticed. More interested in basketball than academics, Adam is an average student, earning high Cs and low Bs. A cool kid with a social group that isn't anywhere to be found on the Science Olympiad team, you can see his wall building as the seconds tick down to the start of class. The bell rings, and I welcome the class to this year's Science Olympiad elective. At the mere mention of the competition, I am greeted with applause and cheers by the rest of the eager students.

The roll of Adam's eyes says it all.

Fast-forward to the Science Olympiad competition awards ceremony. The team of students squirms anxiously in their auditorium seats as they wait to hear the top six teams from each competition. As the announcement for the water bottle rocket competition goes from sixth place, to fifth place, to fourth place, the tension builds. The top three teams are asked to come up to the stage to accept their prizes and medals. Assuming the odds are against us, I glance down at my program to see what event is coming up next. Suddenly, my entire team starts screaming as the judges announce that Adam and his partner have just won second place. Grinning from ear to ear, Adam jumps up, high-fiving his teammates as he races toward the stage to grab his award. Still smiling, he proudly accepts the medal and hangs it around his neck … an accessory he will wear to school the next day to show off to his basketball friends.

THE MORAL

So how did I help Adam change his tune and jump on board? It wasn't with candy, extra credit, or a secret teaching method buried within my science pedagogy textbooks. It was actually quite simple: I took time to get to know him.

In a world where teachers are often asked to teach 150+ students, it's easy for teachers to start generalizing that all students are the same. Although pedagogy such as Bloom's taxonomy and differentiated lessons teach the importance of meeting the needs of each individual student, the reality of teaching often makes this difficult. In fact, teachers are often so caught up dealing with the high- and low-end students, those in the middle often go by with little or no attention.

I know what you are thinking: "Do you really expect me to have the time to get to know every single student I teach? That would take me more hours than there are in the day. As a new teacher, I have lessons to plan, procedures to learn, supplies to buy, and papers to grade (let alone have time to get to actually eat lunch). Maybe I'll be able to do this once I have a few years under my belt, but as of right now … not so much."

And while it's true that it can be difficult to get to know each of your students while juggling so many balls during your first few years, it can be done. Aside from taking advantage of your time in the classroom, some of the best opportunities to get to know students exist *beyond* the classroom.

STEPS FOR SUCCESS

1. COACH.

Taking on a role of mentoring and leadership outside of the academic arena can give you insights into your students that no amount of side conversations in the hall can compete with. All of a sudden you are not there as a science teacher with facts to teach them or lab skills to impart. Instead, you are there to support them in a completely new field void of grades and tests. This lack of academic assessment allows them to get to know you in a world outside of being a teacher. And since you are no longer evaluating them, they are much more open to talking with you about issues that are relevant in their lives. These are issues students want to talk about with their teachers, but the only time they get to see you is during structured (and often hectic) hours of the school day.

2. BE A FACULTY MODERATOR OR ASSISTANT COACH.

But what if you aren't athletic or confident enough to be a head coach? You're in luck: Those people need assistant coaches! This gives you all the benefits of the coaching role, without as much of the planning or knowledge of the sport. I have coached volleyball, basketball, and soccer, despite the fact that the last time I played some of those sports was in my *own* middle school PE class. Being an assistant takes away much of the stress of coaching, but it still allows you to make connections with your students.

3. BE AN ACADEMIC COACH.

Have a fear of tennis shoes? No problem. Instead, take on the role of an academic coach. Whether it's the debate club, science fair, or FIRST LEGO Robotics, there are many options (some even science-related) that allow you the same benefits of athletic coaching. These clubs are often electives that meet before, during, or after school, and they are often more intimate than the classroom environment. As a bonus, students are generally there by choice, thus creating a common bond between you and your students. Overseeing a club is not only a great way to share a hobby or personal passion with your new students, but it also allows them the opportunity get to know you better. As a result, they will often be willing to share a bit more of themselves with you. This relationship will then transcend into the academic classroom as well.

4. CREATE A COMMUNITY OUTREACH EVENT.

If coaching is too large of a time commitment, you can gain many of the same benefits from conducting community outreach (keeping in mind that your students are your community). Possible activities include:

- Plan a stargazing night with your local astronomy club for students. This is a great opportunity for them to learn about constellations and high-powered telescopes from a local expert.
- Organize a rock-climbing night. Hold a brief session discussing the physics behind rock climbing (highlighting friction, simple machines, force diagrams, and that ever-present gravity), and then let them put it into action!

- Arrange for a weekend gathering at a local community garden. Besides improving the garden, you can talk about photosynthesis, limited resources, composting, or symbiosis.

- Plan a night out at the movies. Show a movie in your classroom or the school auditorium that has a science-theme related to your studies like *Gattaca* for genetics, *Osmosis Jones* for cellular biology, or even the latest releases from NOVA, National Geographic, or Discovery Channel (but I'd encourage you to start with mainstream films to create buy-in first).

Check with your school and local community to explore options in locations, transportation, field trip guidelines, and rules for showing films. Although this is often information you can get from a mentor or fellow faculty member, it's always a good idea to clear it with your administrator before promoting the event with students and parents. Keep in mind that making events "optional" often cuts down some of the red tape, and it also allows the group to be somewhat smaller (allowing for more opportunities for conversations with individuals or small groups).

While these opportunities are "one-hit wonders," they still provide opportunities to spend time with kids in a setting that isn't purely academic. Each idea contains a social element to the gathering, and gives students an opportunity to see you trying to weed a garden or choking on your popcorn at a funny moment. By putting yourself in a setting where you aren't viewed as the "all-knowing" teacher, students feel more comfortable conversing with you both in and out of the classroom. This, in turn, leads to a better classroom experience for both you and your students.

WHAT DOES SUCCESS LOOK LIKE?

What you gain from taking on these extra roles in the lives of your students has an exponential payoff in your classrooms. Suddenly, you have students who view you not only as a teacher, but as a person as well. It is clear that if you take the time to both listen to their stories *and* share some of your own, you will form a more solid relationship with your students.

And with that relationship comes trust.

Trust to try a new skill such as gel electrophoresis. Trust that your wacky introduction is going to teach them something amazing. Trust that you really want them to succeed. And this trust translates into students having buy-in for what you are asking them to do in the classroom. If they have seen you taking risks outside of the classroom, then they will be willing to take chances within it.

Adam's story is a great example of this. Here was a student who just needed an opportunity to build a relationship. Within a week of having him on the Science Olympiad team, I learned about his passion for sports, about his family life, and about his feelings about being labeled a "struggling student" ... and that was just the first week! The more he shared, the more engaged he became with the team and in his science class. His participation in my science class skyrocketed, and he was often one of the first students in the room and one of the last students out. He was willing to take the lead with his peers on science labs, and often drew connections between what we were working on for the competition with what we were doing in class. In fact, he was actually caught in another class creating a geometric Science Olympiad sign using graph paper that still hangs in my classroom today.

And the best part was that these changes weren't just happening with Adam. *All* of the students on the Science Olympiad team, whether they won medals or not, had a new heightened interest in my science classroom. Through our shared experiences, I had gained their trust, and because of this, they were willing to take leaps of faith when I pushed them beyond their comfort levels. Additionally, these students guided classmates who weren't on the team down that path as well. These students helped turn the tides in their classrooms by leading discussions and showing respect for what others had to say. Because of this, many of the classroom management issues that I had battled as a new teacher, including low participation and negative comments, were eliminated. As a result, I was able to take more chances in my lessons. The confidence I had gained through these opportunities provided me the chance to try things that I had never dreamed I would be able to implement in my classes based on the early struggles I encountered as a new teacher.

Is it impossible to establish similar relationships and outcomes while in the classroom? Not at all. Many successful teachers create ways to use their curriculum to make these connections while teaching science content and skills. The reason we encourage you to think of options outside of the classroom (in addition to those within) is because it's an easy model for new teachers, with little-to-no experience, to replicate. Additionally, it's an approach rarely taught in teacher preparation, and therefore warrants introduction and explanation. Like all things in education, there is no "one-size-fits-all" model that works for every teacher and student. The best thing to do is to look at your skills and interests, and consider all of the available options to find an approach that works for you, your students, and your community.

Coaching and community outreach has a payoff in the classroom that far exceeds the time commitment required. Whether it's using a talent you have for sports, the arts, a hobby, or just a fun idea you think your students might enjoy, it is worth your time as a new teacher to extend yourself into this arena. The relationships you build will change your classroom into the one you dreamed of when you started down the path to becoming a science educator.

RESOURCES FOR MORE INFORMATION

Yager, R. E., and J. Falk, eds. 2007. *Exemplary science in informal education settings: Standards-based success stories.* Arlington, VA: NSTA Press.

Science education doesn't stop at the schoolhouse door. *Exemplary Science in Informal Education Settings* shows real-world examples of how science education reform has taken hold in museums, science centers, zoos, and aquariums as well as on television, radio, and the internet.

This essay collection—the fifth volume in the *Exemplary Science Monograph Series*—features 17 informal education programs that were judged to be most successful at increasing participants' learning. The programs demonstrate how the Standards can be used to inform and improve science education in a wide range of settings and with learners ranging from preschoolers to older adults.

Froschauer, L., ed. 2008. *Science beyond the classroom.* Arlington, VA: NSTA Press.

Taking science education beyond the classroom provides learning opportunities and experiences for students that just aren't available within school walls, and *Science Beyond the Classroom* has a wealth of ideas on how to do it successfully. These carefully selected articles from the NSTA journals *Science Scope* and *Science and Children* were gathered into a compendium because of the value of informal science education in providing access to those experiences and in tapping into student interests.

Science Beyond the Classroom provides an overview of information and ideas—many of them include step-by-step, teacher-tested instructions and guidelines—that can be easily modified and adapted by teachers and others—scout leaders, club sponsors, parents, and home schoolers among them—who want to nurture enthusiasm for science. An introduction providing background and ideas for using the articles and a list of additional articles and websites leads each of the five sections:

- Extending Science Learning Beyond the Curriculum: Projects and Challenges
- Extending Science Learning Beyond the School Building Walls: Using Local Sites
- Extending Science Learning Beyond the School Day: Clubs, Camps and Science Expositions

- Extending Science Learning to the Family: Take-Home Projects and Family Science Events
- Extending Science Learning to Informal Institutions: Museums, Zoos, and Other Field Trips

With its potential for engaging student interest, this book aims also at encouraging more students to enter academic fields of science, technology, engineering, and mathematics.

Russell, H. R. 1998. *Ten-minute field trips, 3rd edition.* Arlington, VA: NSTA Press.

You don't have to go far to get science out of the classroom. An NSTA bestseller, this book is ideal for teachers in all school environments—urban, suburban, or rural. Renowned educator Helen Ross Russell describes more than 200 short, close-to-home field trips that explore new dimensions of familiar spaces and objects. Brick walls, rock outcrops, lawns, broken pavement, weeds, and trees are all targets for exploration.

Each topic section (plants, animals, interdependence of living things, physical science, Earth science, and ecology) includes pretrip classroom activities, teacher preparation, and a list of trip possibilities. For urban areas, a special cross-referenced list of field trips for hard-topped school grounds is included.

Stebbins, R. 2011. *Connecting with nature: A naturalist's perspective.* Arlington, VA: NSTA Press.

An irresistible story of how one child fell in love with nature and your students can, too. Taking what he calls "a nature-centered worldview," author Robert Stebbins blends activities, examples, and stories with his perspectives on the importance of dealing objectively yet compassionately with social and environmental problems. As thought provoking as it is charming, *Connecting With Nature* includes

- discussions of "ecological illiteracy" and the impediments that keep people, young and old, from bonding with nature;
- recommendations for establishing a nature-centered educational program and encouraging interest in nature at home;
- advice on doing accurate observations and field reports and understanding natural selection; and

- a captivating array of activities to capture the attention of students of all ages: imitating animal sounds, quieting lizards, tracking animals, photographing birds, and playing hide and seek with owl calls.

Even a quick glance through *Connecting With Nature* will make you wish you could give your students the joy of a day in the hills with the author. Failing that, you can use his book to instill a love of nature in your students—and rekindle it in yourself.

Robertson, W. C. 2001. *Building successful partnerships: Community connections for science education.* Arlington, VA: NSTA Press.

No single educator can help children learn all they need to become scientifically literate. Resources are all around us—not only in traditional science classrooms and laboratories, but also in gardens, nature centers, parks, youth programs, museums, and on television and radio. *Community Connections for Science Education, Volume I: Building Successful Partnerships* offers advice on how to select community resource partners, set joint learning goals, improve pre– and post–field trip activities, instruct students in field trip safety and etiquette, and much more.

This book was developed by the National Science Teachers Association and the National Park Foundation.

Lewis, M. 2008. *Coach: Lessons on the game of life.* New York: W. W. Norton.

There was a turning point in Michael Lewis's life, in a baseball game when he was 14 years old. The irascible and often terrifying Coach Fitz put the ball in his hand with the game on the line and managed to convey such confident trust in Lewis's ability that the boy had no choice but to live up to it. "I didn't have words for it then, but I do now: I am about to show the world, and myself, what I can do."

The coach's message was not simply about winning, but about self-respect, sacrifice, courage, and endurance. In some ways, and even now, 30 years later, Lewis still finds himself trying to measure up to what Coach Fitz expected of him.

Intel International Science and Engineering Fair (*www.intel.com/content/www/us/en/education/competitions/international-science-and-engineering-fair.html*)

The Intel International Science and Engineering Fair (Intel ISEF), a program of Society for Science & the Public, is the world's largest precollege science fair competition. Each year, approximately 7 million high school students around the globe develop original research projects and present their work at local science fairs with the hope of winning. Those who do, progress to regional, state, and national competitions. Ultimately, the select few—1,500 promising young innovators—are invited to participate in Intel ISEF. At this weeklong celebration of science, technology, engineering, and math, students share ideas, showcase cutting-edge research, and compete for more than 3 million dollars in awards and scholarships.

FIRST LEGO League (*www.firstlegoleague.org*)

The mission of FIRST LEGO League is to inspire young people to be science and technology leaders, by engaging them in exciting mentor-based programs that build science, engineering and technology skills; that inspire innovation; and that foster well-rounded life capabilities including self-confidence, communication, and leadership.

Dean Kamen is an inventor, entrepreneur, and tireless advocate for science and technology. His passion and determination to help young people discover the excitement and rewards of science and technology are the cornerstones of *FIRST* (For Inspiration and Recognition of Science and Technology).

FIRST was founded in 1989 to inspire young people's interest and participation in science and technology. Based in Manchester, New Hampshire, the 501(c)(3) not-for-profit public charity designs accessible, innovative programs that motivate young people to pursue education and career opportunities in science, technology, engineering, and math, while building self-confidence, knowledge, and life skills.

Science Olympiad (*http://soinc.org*)

Science Olympiad competitions are like academic track meets, consisting of a series of 23 team events in each division (Division B is middle school; Division C is high school). Each year, a portion of the events are rotated to reflect the ever-changing nature of genetics, Earth science, chemistry, anatomy, physics, geology, mechanical engineering, and technology. By combining events from all disciplines, Science Olympiad encourages a wide cross section of students to get involved. Emphasis is placed on active, hands-on group participation. Through Science Olympiad, students, teachers, parents, principals, and business leaders bond together and work toward a shared goal.

Teamwork is a required skill in most scientific careers today, and Science Olympiad encourages group learning by designing events that forge alliances. In Elevated

Bridge, an engineering whiz and a kid from woodshop can become gold medalists. Similarly, a talented builder and a student with a good science vocabulary can excel in Write It Do It, one of Science Olympiad's most popular events.

The culmination of more than 240 regional and state tournaments is the Science Olympiad National Tournament, held at a different university every year. This rotating system gives kids a chance to visit new parts of the country, and tour colleges they might consider for their undergraduate studies, and provides a memorable experience to last a lifetime.

U.S. Department of Energy National Science Bowl (*http://science.energy.gov/nsb*)

Launched in 1991, the National Science Bowl (NSB) is a highly competitive science education and academic event among teams of high school and middle school students who compete in a fast-paced verbal forum to solve technical problems and answer questions in all branches of science and math. Each team is composed of four students, one alternate student, and a coach. Regional and national events encourage student involvement in math and science activities of importance to the Department of Energy and the Nation.

The National Science Bowl for Middle School Students was started in 2002 and includes two types of competitions: an academic math and science competition and a model car race. The car race provides the students with a "hands-on" science and engineering experience where the teams design, build, and race their model cars.

Regional science bowl championship teams receive an all-expenses-paid trip to compete at the national event. High school and middle school teams travel to Washington, D.C. in May. The national events are several days of science activities, sightseeing, and competitions. Teams enjoy the entire science bowl experience and take home many prizes. There are cutting-edge science seminars and hands-on science activities.

Google Science Fair (*www.google.com/intl/en/events/sciencefair/index.html*)

The Google Science Fair challenges students age 13–18 to carry out a scientific investigation into a real-world problem or issue that interests them. The competition asks them to carry the investigation forward through rigorous experimentation, recording, and conclusions.

Students compete with peers in their age group from all over the world to win scholarships, internships, and once-in-a-lifetime experiences. There will also be a special Science in Action prize, sponsored by Scientific American. This will reward the project with the highest capacity to make a practical difference to the lives of people in a group or community.

The 2012 Google Science Fair is a unique opportunity for young people to engage with the scientific community at large.

NSTA's "Blick on Flicks" (*www.nsta.org/publications/blickonflicks.aspx*)

We all love watching movies. But we also love science. And sometimes the two don't mix! To help us sort the good science from the bad in movies and other visual media, Jacob Clark Blickenstaff, PhD, provides expert commentary, pointing out where the physics is stretched, the chemistry fudged, or the biology twisted on behalf of the story—without losing sight of the fact that movies are meant to entertain. Jacob helps turn "bad science" into teachable science for middle level and high school students. "Blick on Flicks" is a regular column in *NSTA Reports* and a periodic feature of NSTA WebNews. His previously published reviews are available on the NSTA website above.

Math, Engineering, Science Achievement (MESA) (*http://mesausa.org*)

Mathematics, Engineering, Science Achievement (MESA) is nationally recognized for its innovative and effective academic development program. MESA engages thousands of educationally disadvantaged students so they excel in math and science and graduate with math-based degrees. MESA partners with all segments of higher education as well as K–12 institutions.

MESA has a proven track record with over 40 years in producing math-based graduates by providing support such as classes, hands-on competitions, counseling, transfer support and a community environment to students from middle school through four-year college.

Since 1970 MESA has helped students become scientists, engineers and mathematicians, filling an urgent need for qualified technical professionals. Through three branches of MESA, students from all segments of education are served through local centers. MESA partners with industry and all the major educational institutions throughout the United States.

Most MESA students are the first in their families to attend college. Most are low-income and attend (or have attended) low-performing school with few resources. MESA serves students in pre-college through the MESA Schools Program (MSP), community college students through the MESA Community College Program (MCCP), and four-year college level students in the MESA Engineering Program (MEP).

PROFESSIONAL DEVELOPMENT
LEARNING to ALWAYS BE A STUDENT

"If you are going to be a teacher, you should always be a student." This quote has directed my entire career in education, and continues to drive my actions today. I share it with every single new teacher that I meet in the hopes that it translates into experiences and efforts that benefit the teacher and his or her students. The importance of professional development not only impacts you as a new teacher, but it can translate into a culture of growth and change that supports your continued development as you move from new teacher to veteran teacher (and beyond). This chapter highlights how to find professional development, make good use of it, and how to pass it along to others back in your school environment.

THE STORY

"Have you ever thought about going to a conference?" a veteran science teacher asks me as I am frantically rushing around my room preparing for the next class in five minutes.

"Nancy, I am so behind I can't even eat lunch in the faculty room. Chances of me leaving my classroom beyond the standard bathroom break look bleak," I respond as I breathlessly grab star charts for our upcoming discussion on the celestial sphere. "I would love to go to a conference, but maybe in a couple of years, after I have survived the gauntlet that is my current reality."

Me at a conference? This must be a new form of new teacher hazing. There was no way I could go to a conference as a first-year teacher.

Six weeks later, I am sitting on a shuttle from the airport to my conference hotel with Nancy. We are going to share a room at the cheapest hotel in town to cut costs, and I'm incredibly nervous. Not about sharing a room with Nancy (she was also a science teacher, so clearly we were compatible), but worried about what was going to happen in my classroom while I was away.

I had left a four-page substitute plan detailing the progression of activities that needed to take place for my one day away from the classroom. I made sure I laid out the attendance lists with notes about support for specific students. I also made a list of where additional supplies could be found (although I had already set out all of the materials necessary). I even provided optional extension activities just in case my students finished early.

But even with my overpreparation, I was still nervous about what I would come back to find on Monday morning. Could my students survive without me? Would the substitute survive my students? Would my room still be upright when I returned? With questions like these buzzing in my head, one thing suddenly became clear—this was a huge mistake.

THE MORAL

Leaving the classroom is never easy, even if you aren't a new teacher. Whether it's the relationships that you form with your students, the overarching science themes for your class, or even something as simple as knowing where the staplers are, it's hard to hand over your classroom community to another teacher. (See Chapter 11, "Calling in Sick," for more detail on this idea.) But you must conquer this hurdle in order to engage in one of the most important opportunities as a teacher: Attending professional development.

When initially presented with this idea, I found it confusing. Hadn't I learned everything I needed to be a teacher-of-the-year candidate in my student teacher experience? (If you are more than a week into your first year of teaching, you know this isn't the case.) It seemed like this was more of a "teacher vacation" scam rather than a legitimate career-based decision.

What I failed to realize was that being a teacher required me to be a constant learner. This was important for several reasons:

1. It gives you some perspective on what's it's like to be a student again. The longer you teach and stand in the role of "expert," the easier it becomes to not understand why students can't rattle off the characteristics associated with families in the periodic table after you have shown them the trends (twice!). Over time, things become "obvious," but only to the person (you) who has had the opportunity to repeat the lesson. Attending professional development puts you back in the position of learning. Rather than knowing "everything" like you do in class, professional development provides the opportunity to ask questions and discover new ideas in ways that the classroom can't. We are constantly asking students to take chances in their learning, and professional development gives you, as a teacher, the opportunity to take these risks as well.

2. Professional development is inspiring. It helps remind you that you have the incredible opportunity to impact students' ideas and development in science on a daily basis. The insights, skills, lessons, and questions that are shared at professional development can re-ignite your passion for teaching, even in the face of the challenges you experience as a new teacher.

3. Engaging in professional development helps you embrace change. As an educator, it's important to recognize that change is an essential part of the job. You must be open to altering what you do in class as a way to constantly

improve the science education that you are offering students. This includes changing approaches, lessons, teaching strategies, and administrative policies. Embracing change (rather than remaining stagnant) keeps your focus and goals on your students' learning, and you will, over time, be less complacent and resistant to new ideas as they emerge in education.

4. Professional development gives you the opportunity not only to learn new information, but it also allows you the chance to share what you *already* know. As you talk with other teachers, you will find yourself explaining and supporting what you currently do in your science class. Doing so not only forces you to think about the "how" and "why" of your practice, but it also opens the door for other suggestions to be added. Plus, offering your "expertise" is a perfect segue into finding what other teachers do (and *don't* do) in their classrooms.

5. Finally, it's a great place to get amazing ideas and resources for your classroom that are *free*. That's right, there are many agencies out there looking to engage with teachers and support their science education in the classroom. And with more than just learning, many opportunities provide free resources, lesson plans, and even classroom materials for participating teachers. Now, don't show up to each opportunity expecting "Oprah's Favorite Things for Science Teachers" and to walk away with a new hybrid vehicle, but it's a great place to get materials if you are in short supply.

STEPS FOR SUCCESS

1. ENGAGE IN PROFESSIONAL DEVELOPMENT.

This is a required element of our profession, just as it is in the medical field, legal practice, and business world. As an educator, constant learning will help you strive to be the best teacher that you can be for your students.

Where can you find this professional development? There are many resources that offer outstanding professional development that caters to different topics and skills. Those groups include:

- ***School and district professional development.*** Many schools and districts offer professional development programming for their teachers. These are often site-specific to address needs that have been specifically evaluated in your community. Keep in mind, these are often about students and/or teaching, and they may or may not be specific to science. You can also look to other district and state-based resources available through your state Office of Education.

- *Teacher association groups such as NSTA, NSLEA, and ASCD.* When looking for professional development that is specific to science teachers, these groups are outstanding resources. They offer many onsite professional development experiences, including conferences and workshops. Additionally, they also offer supplemental resources in the form of print materials and webinars. They will have cutting-edge information and research about the best practices for science teaching, as well as exposure to curriculum and practices from outside of your current school and state community.

- *Listservs for specific content areas, such as American Institute of Physics, American Chemical Society, and the Howard Hughes Medical Institute.* These groups offer content-specific resources and professional development that can help you solidify content and skills associated with your specific science discipline. Many of these organizations offer communication venues such as blogs, wikis, or e-mail Listservs so members can communicate with other teachers to seek specific support and share ideas.

- *Informal education sites such as libraries, museums, zoos, and art institutes.* Informal education sites within your area are a great place to seek out local professional development. Aside from being knowledgeable about education practices, they are also great places to find student resources and programming to support and extend the science learning beyond your classroom and into the community. These sites can also be a great place to engage with other science educators who are within your state, but outside of your school and district.

- *Universities and community colleges.* Most teachers think that once they complete their degree, their time on a college or university campus is complete. But these institutions are great places for continuing education. Aside from the teacher preparation courses, many of them offer learning courses onsite or online for current classroom teachers.

- *Research institutions, many of which are required to do outreach for K–12.* Whether at a college of science or a specific research institute, these can be great sites to connect with research scientists and projects. These can be beneficial because they will increase your own science knowledge, and help you find cutting-edge materials and connections to inspire your students in the classroom. Most research science is funded by grants, which require elements of community outreach. Making personal connections with these scientists is a great way to get you and your students involved with their projects.

One additional item to consider is to look at *who* is hosting a PD workshop/event:

- Teachers often offer the best sessions, as they usually present curriculum and pedagogy that has been successful in their classrooms. This is a great place to get vetted curriculum, but be prepared to keep an open mind. You will most likely need to make tweaks to the presented ideas to meet the needs of your students, community, and resources.

- Educational company sessions are most likely trying to sell you their product, but they also often provide free samples to take back to your classroom. If you are in a position where you are looking to buy new materials, this is the best way to learn about what items make sense for your classroom. Even if you aren't in a position to by new equipment right away, learning about what is available can be helpful for planning your future budget requests.

- Government agency sessions give away *lots* of resources, including curriculum, posters, and classroom materials. Because they are sponsored by tax dollars, these are all free. There are many free national resources through agencies such as NASA, the National Science Foundation, and the National Institute of Health. Agencies such as these also can be a resource for additional teacher professional development and programming.

2. IT'S OK TO "SHOP" SESSIONS.

Have a couple of options that you are interested in attending, and start with your first choice. If you realize one isn't going to meet your needs, move on to another. Your time is valuable, so you want to spend it getting resources and knowledge that make sense for your curriculum, resources, and community.

3. WHEN YOU COME BACK TO YOUR SCHOOL, EDUCATE OTHERS!

Ask to make a short presentation at a faculty meeting or as a part of the next PD day. There is a *huge* misconception out there, especially amongst new teachers, that the people who do the professional development presentations create their own curriculum. NOT TRUE! (Evidence: Can you think of a lesson idea that you have seen presented by more than one person? It's not serendipity that led them both to the exact same conclusion. The more likely scenario is that *they* heard it from another person, have since used it with success, and are now passing it along.) Sharing what you learned helps to demonstrate your passion for learning, and it can lead to collaboration with colleagues, future professional development experiences, and leadership opportunities within your school.

WHAT DOES SUCCESS LOOK LIKE?

By spending one day away from the classroom, I gained ideas and materials to last me almost six full weeks. These were tested resources that made my "Google search" approaches pale in comparison. The time spent engaged in professional development helped me offer a better experience to my students, and not just for *that* year, but for years to come.

The time with Nancy also led to collaboration, where we took the opportunity to pilot some of our new material by switching grade levels for the day. The professional development sparked conversations about what was happening in our classroom, and it led to the exchanging of best practices. Although Nancy was my school neighbor, we rarely had an opportunity to talk at length about how we both approached science. It was incredibly helpful to hear about the structure and skills she was promoting in her class so that I could build upon those with my students in eighth grade. Professional development is a great time to have conversations that go beyond the traditional "You will never believe what a student did in my class today …" chats held in the faculty lounge.

Finally, professional development is what allowed me to not only demonstrate my commitment to teaching, it also opened additional doors for growth and leadership within the science community. By sharing my learning from these experiences, I emerged as a resource for other teachers, and as a result, I have developed more connections within my school, state, and national community.

The difference between the learning experiences I was able to offer in my first year as opposed to my third was monumental. The improvements emerged not just from having first hand classroom experience, but from continuing to pursue professional development.

Looking back, I am thankful that Nancy dragged me to that first conference, and I encourage all of you to search out these opportunities as teachers. I promise that your students will survive without you for a day, and even though they might not say it, they will be all the more grateful for your improvement.

RESOURCES FOR MORE INFORMATION

NSTA Professional Development (*www.nsta.org/pd*)

This website provides a list of national opportunities, including:

- *NSTA Learning Center.* If you want a better understanding of what you teach (the science content in your subject areas) and how to teach it (techniques to help your students learn) the NSTA Learning Center is the place for you. We developed this 24/7, electronic professional development website with your classroom needs and busy schedule in mind.

- *NSTA Web Seminars.* NSTA web seminars are 90-minute, live professional development experiences that use online learning technologies to allow participants to interact with nationally acclaimed experts; NSTA Press authors; and scientists, engineers, and education specialists from NSTA government partners such as NASA, NOAA, FDA, NSF, and the National Science Digital Library.

- *Science Matters Network.* The mission of the Science Matters network is to end the isolation of classroom science teachers and to provide them with professional development opportunities and science teaching resources. The goals of the program are to identify a Point of Contact for standards-based science teaching and learning in every school; establish an electronic network of science educators; and create an infrastructure of national and state partners who are advocates for effective science teaching and learning.

Koba, S. B., B. S. Wojnowski, and R. E. Yager, eds. 2013. *Exemplary science: Best practices in professional development, revised second edition.* Arlington, VA: NSTA Press.

This collection of 16 essays is ideal for staff development providers (university faculty, district supervisors, lead teachers, and principals) as well as preservice science methods instructors. Each essay describes a specific program designed to train current or future teachers to carry out the constructivist, inquiry-based approach of the Standards. These professional development programs are in universities and school districts around the country—from Florida to Alaska, in rural and urban areas, and in contexts ranging from summer institutes to onsite support programs.

In addition to describing how the program works, each essay also provides evidence of effectiveness—how teachers grow more confident using inquiry approaches, in integrating technology into the classrooms, collaborating with their colleagues and local resource persons, and developing as science education leaders in their schools and districts.

Rhoton, J., and P. Bowers. 2001. *Issues in science education: Professional development leadership and the diverse learner.* Arlington, VA: NSTA Press.

Make science teaching better for every student. Help learners from different backgrounds—and with different learning styles—by developing new skills, resources, and knowledge. This book discusses the ways in which professional development can help you handle equity and diversity issues in the classroom. The following topics are explored:

- How professional development can help teachers motivate and increase participation by women and minorities in science
- Using professional development to promote change
- Professional development's role in leadership development and reform

CALLING IN SICK

WHAT to DO WHEN YOU ARE OUT OF THE CLASSROOM

Whether you have been teaching for 1 year or 50, no one likes to miss class. But circumstances will arise throughout the school year that will require you to turn your beloved students over to someone else. And while the thought of this might be scary, the following chapter will help ease your "I can't miss my class!" phobia.

THE STORY

It's 3 a.m. on a Tuesday in January, and I am wedged between the bathtub and towel rack on the bathroom floor. It's been a horrendous evening of illness where I have reflected on the error of not replacing the hand sanitizer in my classroom when it ran out a few weeks ago. Whether it's food poisoning, a virus, or a bacteria, I am beyond the point of caring, and I am starting to wish that I paid more attention to my urgent care options in the start of school year health insurance meeting. Despite my current state of disease, the *only* thing on my mind is how I am going to get dressed and prepared for the lab setup for my class that starts in less than five hours. The song "Should I Stay or Should I Go?" starts vamping in my head as I lunge toward the toilet bowl.

THE MORAL

Teaching requires a teacher be present, which often explains why educators have such a difficult time being out of the classroom. However, whether it's to pursue professional learning of content or pedagogy, or a dodgy meal with gas station sushi, most teachers will find themselves in a situation where they have to turn over their classroom to a substitute.

As a new teacher, this was so alarming that I often dragged myself into the classroom when I really should have been in quarantine. This is a bad idea. Not only do you expose your students and colleagues to germs, but no one wants to see their teacher in even worse condition than when they stay up until midnight grading lab books. Not only do you put your students at potential risk for spreading the illness, but you also extend your illness and impair the impact you have on your classroom.

Additionally, this book makes a strong case for attending professional development. Many of these opportunities also come during a school day, so it's important that you have a plan in place to be able to take advantage of these opportunities when they have potential for improving your teaching for your students.

Recognizing that you *will* have moments that you absolutely cannot be in the classroom, check out the following steps to success so that you can make the best decision for your health or learning when the time arises.

STEPS FOR SUCCESS

1. HAVE A SUBSTITUTE PLAN ON FILE WITH THE OFFICE.

Since you never know when you might be out, this step is an important one to take at the start of the year! One suggestion for this lesson is to have it focus heavily on a science skill or practice (such as asking a question, analyzing and interpreting data, engaging in argument from evidence, and so on) as opposed to content. This way you aren't in a position to be changing the lesson every time you change units. Also, having materials and supplies set aside in the office or your classroom for easy access for the substitute is also highly recommended. Finally, it might not be a bad idea to have a backup plan as well, just in case the first one goes a little faster than you planned.

Although most schools have policies and procedures to support a substitute, it can be helpful to include a student roster and list of classroom policies with the lesson plan as well.

2. USE YOUR RESOURCES ... INCLUDING YOUR STUDENTS.

Students can serve key leadership roles in the classroom, especially during a planned absence. Targeting a few students and walking them through the plans before the day that you are planning to be absent can often make the difference between a substitute plan being implemented versus coming back to find out that they just watched the NCAA basketball tournament on the projector all day. (And yes, that is a true story.) Students can serve as great peer leaders, especially if they are identified to the substitute. In conjunction with this, the students appreciate the trust and recognition of their leadership in the classroom. This is a great way to give kudos to students who have demonstrated excellence in the learning environment.

3. USE TECHNOLOGY TO YOUR ADVANTAGE.

Although there will be certain instances when you should *not* be available via e-mail (such as in the story at the beginning of this chapter), the technology age has enabled teachers to stay in touch with their classrooms even when not in the same location. Whether it's a morning Skype to get your class on track, or just leaving your e-mail address and letting students know that you hope to hear any questions they have during the day, this additional support can be a valuable resource (if it's feasible with your schedule).

4. KNOW YOUR SUBS.

Meet substitutes as they come to the school for other faculty to find out which substitutes are preferred within the community. You will quickly learn that all subs are *not* equal... trust me! Once you have a list of preferred substitutes, share that with the school. If you are not able to arrange for your own substitute, it's good that those who do, know which individuals to contact first.

5. LET STUDENTS DO SOME CREATIVE WRITING.

Allow your students to show you what they already know through literacy activities that go beyond traditional nonfiction. There are lots of creative ways for students to reinforce the skills and topics that you have recently covered in class. Have them create a comic illustrating lab safety rules. Let them write a creative story where they have to incorporate key vocabulary terms. Give them the chance to create an imaginary conversation where two famous scientists (Einstein, Curie, Tesla, Goodall, and so on) argue with each other about why they are the most influential scientist ever. Students love the chance to be creative, and any of these activities will serve as a great discussion activity (or an informal presentation) once you return.

6. A LITTLE NOTE GOES A LONG WAY.

During an unplanned absence (such as being sick), if possible write a note on the board (or have the sub read one aloud) explaining your situation. Let the students know that they are expected to behave the same for the sub as when you are there. While this might sound simple, something as small as saying, "I'm sick, and I need your help to keep things going" can go a long way with students.

7. LET YOUR FELLOW TEACHERS KNOW OF YOUR ABSENCE.

Not every school community has a system for communicating that you will be out to other teachers. A simple e-mail is often helpful to let your colleagues and mentor teacher know that you'd appreciate it if they could check in on your class. It's one more step in the line of defense against finding out they played "roller chair tag" all day (another true story).

WHAT DOES SUCCESS LOOK LIKE?

Having days where you are unable to make it into your classroom, whether for illness, professional development, or personal reasons, is a part of teaching that is often overlooked until it's too late.

And while missing class can feel scary at first, taking a few precautionary steps can go a long way in easing your absence. A well-planned absence (even when it's *un*planned), including a clear substitute plan, student leaders, and a reliable substitute, means that your class will run *almost* as smoothly without you as it does with you. Now will there still be some things that fall through the cracks? Sure. But by taking the above steps, you can limit the issues that arise in your absence.

With a little planning and preparation, it is possible for the students to have a productive learning experience even without your presence. It's shocking, but true! Now trust me, your students would rather have you there just as much as you would rather be there, but it's important to have a contingency plan so that science education can continue in your absence.

And in many cases, you actually return to a hero's welcome with cheers that you are back as they relay tales of struggle and strife in your absence. And nothing really beats an illness like a chorus of students saying "You're BACK!"

RESOURCES FOR MORE INFORMATION

When looking for good activities, I often used some of my own creation, but also some from curriculum resources that I had accumulated as a teacher. These books often contained activities *and* all of the instruction for how to pull them off, which made my substitute plan less challenging and time-consuming. Here are a few examples of resources:

Haysom, J., and M. Bowen. 2011. *Predict, observe, explain: Activities enhancing scientific understanding.* Arlington, VA: NSTA Press.

John Haysom and Michael Bowen provide middle and high school science teachers with more than 100 student activities to help the students develop their understanding of scientific concepts. The powerful Predict, Observe, Explain (POE) strategy, field-tested by hundreds of teachers, is designed to foster student inquiry and challenge existing conceptions that students bring to the classroom.

The 15 chapters cover topics such as force and motion, temperature and heat, light, chemical change, and life processes in plants. Lessons include worksheets, scientific explanations of the concepts being studied, summaries of student responses during the field tests, synopses of research findings, and lists of necessary materials. In *Predict, Observe, Explain*, Haysom and Bowen make it easy for novice and experienced teachers alike to incorporate a teaching method that helps students understand—and even enjoy—science and learning.

Lord, T. T., and H. J. Travis. *Schoolyard science: 101 easy and inexpensive activities*. Arlington, VA: NSTA Press.

With 101 easy and inexpensive activities to do on school grounds, *Schoolyard Science* can help students develop their observation and inquiry skills as well as an appreciation of their outdoor environment. Covering topics such as lower plants, gardens, insects and other invertebrates, energy, and Earth science, Thomas Lord and Holly Travis provide activities that will help teachers become more comfortable with incorporating the outdoors into their curriculum. The activities have been tested successfully in K–12 classrooms, youth camp programs, and science education classes in teacher preparation programs, so teachers can feel confident when using this book in their classrooms.

The book's teamwork focus will allow students to improve their critical-thinking skills and ability to work with other students. Each activity includes a list of the relevant standards and the suggested grade levels; however, the activities can be adapted to other grades as well, allowing teachers to think outside the box. The activities mostly make use of easily accessible materials, but Lord and Travis note any non-school-yard materials that will be needed in a particular activity. Engaging and thoughtful activities make *Schoolyard Science* a great starting point for teachers as they inspire students to appreciate learning in their own school yard.

Slesnick, I. 2004. *Clones, cats, and chemicals: Thinking scientifically about controversial issues*. Arlington, VA: NSTA Press.

Does human cloning present a threat or an opportunity? Do common cats constitute a major threat to wildlife? Will the development of new chemical and biological weapons deter war or lead to it? If you want students to think—really think—about the science behind some of today's toughest controversies, this book will give you the facts and the framework to provoke fascinating debates.

Clones, Cats, and Chemicals examines 10 dilemmas from the fields of biology, chemistry, physics, Earth science, technology, and mathematics and helps you challenge students to confront scientific and social problems that offer few black-and-white solutions. Each question is presented as a two-part unit: concise scientific background with possible resolutions and a reference list for further teacher reading, and a reproducible essay, questions, and activities to guide students in debating and decision making.

Weber, T., ed. 1995. *Quantum quandaries.* Arlington, VA: NSTA Press.

For extra credit or just for the fun of it—why not try a brainteaser? This collection brings together the first 100 brainteasers from *Quantum* magazine, published by the National Science Teachers Association in collaboration with the Russian magazine *Kvant*. Through its pages, you'll find number rebuses, geometry ticklers, logic puzzles, and quirky questions with a physics twist. Students and teachers alike will enjoy these fun quandaries.

Eisenkraft, A., L. D. Kirkpatrick, and T. Bunk. *Quantoons: Metaphysical illustrations.* Arlington, VA: NSTA Press.

Do you crave both physics problems and captivating illustrations? If your answer is "yes," look no further! *Quantoons* combines challenging problems and provocative quotes with intricate drawings that mix Isaac Newton and Marie Antoinette with Romeo, Juliet, and Einstein. The book is a compilation of 58 contest problems that ran between 1991 and 2001 in *Quantum* magazine; a collaboration between U.S. and Russian scientists that was published by NSTA.

Chapter 12

INTERDISCIPLINARY WORK

DON'T BE AFRAID OF THE GOOD TEACHERS

This chapter focuses on how to collaborate with other teachers in your classroom. It provides a guide for identifying teachers to work with, as well as good strategies for building interdisciplinary units with your colleagues. These steps can be used to create units that involve math, American Studies, English, art, and physical education. The chapter will also focus on the idea of using good teachers for inspiration as opposed to intimidation.

THE STORY

While cleaning up the classroom and gathering abandoned school supplies at the end of the day, I came across a copy of *Tuesdays With Morrie* underneath one of the student desks. I grabbed it, threw it in the basket with the three graphing calculators and four history textbooks I'd also discovered, and headed down the hallway to return the supplies to their correct classrooms.

After listening to the history teacher complain about the recklessness of our students for 10 minutes, I popped into the English room to return the book to Mike, the English teacher.

"Another lost book has made its way into the science classroom," I said as I hand the book back over. "I know we all have to do literacy, but I don't think this one really applies to my genetics unit," I joke, as he hands me several metersticks and markers from the science classroom.

"Spoken like a true science teacher. I'm sure it's hard to recognize a book that doesn't weigh the traditional 20 pounds," Mike responds as a science textbook hits the bottom of my basket with a thud.

"It does seem a bit unfair that they make your books so light, and my books are all so heavy. It must be all of the knowledge," I retort as I head back out of his classroom.

"As someone who claims to teach about the world around you, you may actually like this book. It doesn't have bolded vocabulary words, but you are welcome to take a copy. At the very least, it would give you something to talk about with the students." Mike says as he hands back *Tuesdays With Morrie.*

Two weeks later, I have students measuring their breathing and heart rates while breathing through a drinking straw. We are simulating the impacts of the genetic disease, Amyotrophic Lateral Sclerosis (ALS) on the human body. The experience is providing a context for discussing the heredity patterns associated with the disease and the genetic deletion that has been identified as a cause. It also happens to directly tie in with the student's assigned reading in *Tuesdays With Morrie,* which details the struggle of a man with ALS. The students want to expand the experiment into a

Guinness World Record attempt to see how long they could go with only breathing through a straw. "I don't think so. That doesn't sound like a safe science practice!" I exclaim. "But feel free to ask the English teacher when you get back to his class. He is known for taking chances."

THE MORAL

In the same way that the supplies I found scattered in my classroom after school represented all manner of different content areas, science classrooms provide the perfect breeding grounds for collaboration with other topics. Since we teach "the world around us," it's helpful to draw connections with other topics, be it traditional STEM, the arts, the humanities, or even driver's education. These connections help students to establish relevancy for the science concepts and skills, and it creates an increased investment and interest in the topics by providing a strong context.

Sounds pretty simple, right? Just find a sassy teacher down the hall to give you a book that directly ties into your curriculum, and there you have it: Interdisciplinary lessons in action! In all honesty, this particular incident occurred by chance, but led to future collaborations with English, art, and other courses. Getting the idea to collaborate between science and another discipline is one thing, but putting those ideas into action takes the challenge to a whole new level.

And those challenges often start with finding fellow colleagues who are interested in collaborating. Many teachers are already committed to their curriculum and lessons, and therefore struggle to see how collaboration would fit into their classroom.

The best place to start is by looking for the "great" teachers. These are the teachers who current students rave about, former students come back to visit, and they get the best gifts from families come holiday season.

As a new teacher, these "all-star" teachers can often be intimidating. But keep in mind that they are often great because they are always looking for new and exciting ways to engage with students. The opportunity to collaborate has that potential, and most of the great ones will jump at the chance to team with you. (Note: This means being a part of the planning process. Teachers aren't always excited to implement a plan that someone else has already created.) And while finding teachers to collaborate with might not be easy, it is definitely worth it. The best teachers *want* to collaborate, so don't hesitate to take that first step.

The payoff for your effort is that you not only help your students get a better understanding of the material, but you gain insight into a wide range of teaching strategies from those you collaborate with. For example, I learned how to use literature in my classroom when building my experience around *Tuesdays With Morrie.*

This wasn't training to be an English teacher by discussing plot, character development, and overarching themes, but rather centered on drawing connections between the story and students' lives as a way to create relevancy for science concepts. Things like asking how foreshadowing is similar to making a hypothesis or having students compare and contrast the elements of journal writing with science lab write-ups were two simple connections that didn't require me to have a degree in English.

The logic here is the more students get exposed to a topic, the more likely they are to understand that subject. Collaboration provides this opportunity for a deeper understanding by shedding light on an issue from two different subject matters. An example of this centers on my early struggles to engage the entire class in physics discussions about light and color. It wasn't until I brought in the art teacher that *all* my students grasped the concept. Whereas the electromagnetic spectrum resonated with some kids, the aesthetic approach from the art teacher helped others to connect. It also provided a great debate about whether white was all colors (science) as opposed to the absence of color (art). The final project of constructing lamps engaged both their scientific and artistic skills to demonstrate understanding in an application.

Plus, beyond the learning that goes with collaboration, it's always fun to freak the kids out by showing them that you know what's going on in their other classes! For example, after collaborating with Mike, I asked him for a copy of his assessment on *Tuesdays With Morrie*. He provided great instructions for students to organize their short answers in a format that had strong literary elements that I could easily translate into my science class. The next science quiz I gave used the *exact* directions used in their English class for their short-answer responses, and I loved watching their faces as they read the directions to the short-answer questions.

"Did you get this from English class?" the students would ask, a puzzled look on their faces.

"How do you know he didn't get it from me?" I would respond with a smile.

Using the same concepts, skills, and instructions across curriculums creates a more cohesive learning environment that actually *improves* the science content. By creating connections with the world around them, students have more opportunities to link what they learn in science to what they do every day.

STEPS FOR SUCCESS

1. COME UP WITH AN IDEA.

The first step toward collaboration involves you finding a topic that links itself between two or more classes. When you are trying this for the first time, I would recommend finding a way to link a topic you already teach to one other subject. Collaboration takes effort, and rather than starting from scratch, using something that you're already familiar with can make this challenge more manageable.

It's also helpful to choose a topic where you have interest and expertise. You are more likely to come up with connections with topics where you have a clear passion and knowledge base. Select something that is exciting to you, and it's easier to sell the idea to the other teachers.

Once you've established a link, create a brief outline illustrating the connections between the subjects. Again, until you've done this a few times, I would recommend keeping these links fairly simple. You aren't trying to repeat the same topics in both classes, but rather are using a common resource to highlight subject-specific ideas that create a meaningful connection for your students. Pick one or two easy connections that you are confident will be successful. Remember, you are not trying to collaborate once … you are trying to build a *yearly* collaboration. The best way to ensure this is to make it successful, regardless of how small that success might be.

2. FIND THE RIGHT TEACHER.

As mentioned above, a key element to successful collaboration is finding other teachers who want to collaborate. The best way to do this initially is to simply pay attention to how the teachers around you speak and act. Generally speaking, the top teachers are excited about what they do, and they love to talk about the successes their students have. These are the ones you will want to approach.

When you approach them, do so in a way that is inviting and showcases how working together will improve what they do in their class. Remember, collaboration is something that should be beneficial to *both* classes, and if you can't think of a way to help your co-collaborator, then it is not going to be a successful project.

It's often helpful to give them the general idea or resource you'd like to share, then let them develop their own ideas. Mike gave me *Tuesdays With Morrie*, and he let me go and find the science connections. This allowed me to be more excited and invested in the planning. Each teacher is going to be the expert in his or her content, so don't force a full unit plan on your colleague. Instead, let them come up with connections for their class, and then create together.

3. START SMALL.

While you are going to be very excited to get started, I would advise starting small. Find out what works, then build from there. If you dive in too fast and things don't go so smoothly, both you and your co-collaborator will be hesitant to try working together again. Remember, you can always add things to it if you feel the need. The first connection with *Tuesdays With Morrie* was a simple experiment that lasted two days before we launched into the science concepts. It's perfectly fine to hold off on planning a major collaboration project until you have a few smaller collaborations under you belt.

4. HAVE A PLAN.

Once you have a general outline of the interdisciplinary experience, get together with your collaborator and plan out exactly what the collaboration will look like. Potential issues to address include:

- Finding a common language for key terms.
- Will there be any co-teaching involved?
- Are you going to touch upon the same issues, or will both of you do individual concepts around a central idea?
- Are you going to address the topic on the same day? Alternating days?
- Will the topic run the length of the entire unit, or just touch briefly upon certain aspects of it?
- Is there an essential question that can connect the two classes?
- What is the assessment in each class going to look like?
- Could there be one final project that combines elements of both classes?
- Is there potential for a class "swap"? (Switch classrooms so you teach the other subject.)
- How will you assess the success/failure of the unit?

From experience, the more issues you can address prior to starting, the more smoothly the collaboration will go.

5. UNDERSTAND THAT THERE WILL BE BUMPS ALONG THE WAY.

Regardless of how well you plan or how great of a team you and your collaborator are, you need to expect that things won't always go exactly as planned.

And that's OK.

Be flexible in both how you teach and what you expect from the other teacher, and don't get frustrated if things go off course. If this happens, re-examine why it didn't work, and learn from it rather than dwelling on the failure. Similar to your relationship with your mentor, set up a time and format for sharing both successes and struggles with your collaborating teacher. Like everything else in teaching, collaboration becomes easier and better the more you do it. Even if things don't go so well the first time, be open to trying again.

6. DISCUSS, DISCUSS, DISCUSS.

While there is a great deal of preparation that goes into working together, frequent communication should also be happening throughout the interdisciplinary experience.

Daily "touching base" sessions should be occurring between the two of you. These are five-minute discussions covering both the previous and forthcoming lessons. You should address what did and didn't work, as well as making sure you are both on the same page for the upcoming class(es).

In addition, after the collaboration is over, plan on having a longer "debriefing" session. This is where you get a little more in depth about the good, the bad, and the ugly of the experience. This is also the place to make plans to address those items in your future implementation.

Finally, don't forget to involve the students in all of this. Ask them how things are going and if they like what is happening in class. And while I wouldn't base the success or failure completely on what the students say, they generally are a pretty good barometer as to whether a experience is beneficial or not.

7. PRAISE, PRAISE, PRAISE.

Throughout this entire process, please remember to thank your collaborator and to let him or her know that you really appreciate what he or she is doing. Doing something new is challenging and often requires a great deal of work, and a simple sign of appreciation can go a long way.

WHAT DOES SUCCESS LOOKS LIKE?

Successful collaboration isn't a one-time event; it is something that will hopefully become a continual thread throughout your class. The short-answer directions for *all* my assessments now reflect the expectations from Mike's class. This has led to stronger literacy in my science classroom, as I was able to hold higher expectations due to my increased understanding about what was happening in English. This is just one example of how working with another teacher can benefit the science learning experience for your students.

After Mike and I collaborated on our *Tuesday With Morrie* unit, we added another unit the following year, and a third the next. The success with Mike spurred me to reach out to the new art teacher, the health teacher, and the math teacher to create cohesive units. It reached a point where students were surprised when we were "only doing science." This pushed me to find more connections and strive for an interdisciplinary approach with my topics.

It also helped me to find better ways to connect multiple science concepts and skills into my lessons. Rather than teaching concepts individually, I was able to embed multiple science themes, concepts, and science practices into my interdisciplinary approaches. This enriched the science content and learning in the classroom.

The next time you find yourself returning wayward supplies back to their correct classrooms in the building, take a moment to stop and look at the resources available in the other classrooms. You never know when inspiration might hit!

RESOURCES FOR MORE INFORMATION

Linz, E., M. J. Heater, and L. A. Howard. 2011. *Team teaching science: Success for all learners.* Arlington, VA: NSTA Press.

In *Team Teaching Science*, Ed Linz, Mary Jane Heater, and Lori A. Howard demonstrate the truth in the old adage "Two heads are better than one." This guide for developing successful team-teaching partnerships that maximize student learning will help preservice and inservice special education and science teachers in grades K–12, as well as methods professors in science education programs who want to cover special needs issues in their curriculum. Using both research-based practices and personal insight from experienced team teachers, the authors strive to make team teaching beneficial for students and accessible for teachers. Linz, Heater, and Howard provide background information on science teaching and team teaching and, most important, six chapters on how to teach specific science topics and how a co-teaching team can proceed through the school year.

The basic elements of collaboration are introduced, along with chapters on co-teaching strategies to implement in elementary, middle, and high school classrooms. The authors, who have years of co-teaching experience, offer practical advice that teachers can apply to their own classrooms. Teaching a diverse group of students is one challenge teachers will likely encounter in a team-teaching environment; the authors address the difficulties that may arise, as well as issues related to assessment, curriculum, and necessary accommodations and modifications. For those tackling the challenges of team teaching, this book will prove to be a valuable resource for making team teaching a positive experience for both students and teachers.

Harland, D. H. 2011. *STEM student research handbook.* Arlington, VA: NSTA Press.

This comprehensive resource for STEM teachers and students, outlines the various stages of large-scale research projects, enabling teachers to coach their students through the research process. This handbook provides enough detail to embolden all teachers—even those who have never designed an experiment on their own—to support student-researchers through the entire process of conducting experiments. Early chapters—research design, background research, hypothesis writing, and proposal writing—help students design and implement their research projects. Later chapters on descriptive and inferential statistics, as well as graphical representations, help them correctly interpret their data. Finally, the last chapters enable students to effectively communicate their results by writing and documenting a STEM research paper, as well as by preparing for oral and poster presentations. Included are student handouts, checklists, presentation observation sheets, and sample assessment rubrics.

Saul, W., A. Kohnen, A. Newman, and L. Pearce. 2012. *Front-page science: Engaging teens in science literacy.* Arlington, VA: NSTA Press.

Like citizen journalists, your students can get to the heart of science literacy with the "learn by doing" methodology in this innovative book. *Front-Page Science* uses science journalism techniques to help students become better consumers of, and contributors to, a scientifically literate community.

The book is divided into three parts:

- Background information and a rationale for using science journalism techniques

- Concrete advice about how to teach science literacy in this framework—from helping students find story angles to teaching search strategies and interview techniques

- The process of putting together and writing a news story, including how to get students started, help them when they're stalled, and respond to their drafts

A free website provides downloadable lesson plans, teacher suggestions, and a forum for exchanging ideas with others. Like *Front-Page Science*, the website is part of the National Science Foundation–funded Science Literacy Through Science Journalism project. By making full use of these rich resources, you'll teach your students skills that will help them make sense of their world not just now, but also after graduation and for years to come.

Allen, J. 2005. *Tools for teaching content literacy.* Portland, ME: Stenhouse Publishers.

Reading and writing across content areas is emphasized in the standards and on high-stakes tests at the state and national level. As educators seek to incorporate content-area literacy into their teaching, they confront a maze of theories, instructional strategies, and acronyms like REAP and RAFT. Teachers who do work their way through the myriad content reading and writing strategies are discovering not all activities are appropriate for content instruction: Only those with a strong research base meet the high standards expected in classrooms today. Janet Allen developed the ideal support for teachers who want to improve their reading instruction across the curriculum. *Tools for Teaching Content Literacy* is a compact tabbed flipchart designed as a ready reference for content reading and writing instruction. Each of the 33 strategies includes

- a brief description and purpose for each strategy;
- a research base that documents the origin and effectiveness of the strategy;
- graphic organizers to support the lesson;
- classroom vignettes from different grade levels and content areas to illustrate the strategy in use.

The perfect size to slip into a plan book, *Tools* highlights effective instructional strategies and innovative ideas to help you design lessons that meet your students' academic needs as well as content standards. The definitions, descriptions, and research sources also provide a quick reference when implementing state and national standards, designing assessments, writing grants, or evaluating resources for literacy instruction.

Tierney, B., and J. Dorroh. 2004. *How to write to learn science.* Arlington, VA: NSTA Press.

Make science an exhilarating process of discovery! Through a wealth of creative write-to-learn strategies, this book offers inspiring techniques to coax out the reluctant scientists in your classroom. Newly updated and expanded, this NSTA bestseller is a storehouse of practical ideas and examples for use with students at all ability levels. It provides classroom-tested writing activities that you can

- Introduce during the first week of class to build positive attitudes among students toward the subject of science, and toward you;
- Use at different stages in a learning unit and for quick review; and
- Adapt to help students write for different audiences, write to better understand the textbook, and write lab reports, research papers, and essay tests.

Added to this edition is a special section—How Science Portfolio Assessment Can Improve Student Writing—that describes ways portfolios help students focus on their work throughout the year, document science concepts they've mastered or not and serve as powerful assessment tools. There are many books about writing to learn, most authored by education or English professors who focus on theories of writing. This book is different it s full of classroom-tested, pragmatic approaches from high school science teachers who used the ideas to make teaching and learning more creative endeavors. The authors put their own good advice to work, writing in an appealing, personal style to convey teaching concepts and learning goals. As Bob Tierney says, expressive writing is a vehicle for the exhilaration of discovery.

SURVIVING YEAR ONE AND TRANSITIONING INTO THRIVING IN YEAR TWO AND BEYOND

As you finish your rookie season in the world of education, every first-year teacher deserves a hearty congratulation. Everyone will tell you that their first year was the most challenging… what they often omit is that their second year was second on that list of challenging moments as an educator. This chapter supports new teachers as you make the transition between your first few years to develop skills and organization to keep you on the path of improving competency and community impact.

THE STORY

The bell is about to ring and I'm excited, as opposed to the raging case of nausea that had plagued me this time last year. I am ready for the start of my second year as a teacher who has pencils sharpened, chemical glassware sparkling in the clean supply cabinet, and my bicycle wheel ready for the inertia demonstration. I can tell students the names of the other teachers, what time the third period bell rings, and even the page in the student handbook that articulates no gum chewing. Last year I was an immigrant to the school community, but as I begin my second year I am confident that I am now a full-fledged citizen and *veteran* teacher in this community.

My confidence surges as students enter the room and know my name and what class I teach. A few even mention that their sibling was in my science class last year. I hand out their schedules and kick off the school year flashing my knowledge base as I give a new student directions on how to find the PE locker room, talk about the class trip that will occur in a few weeks, and even answer a few questions with confident answers about school policy. I am on cloud nine, and dismiss the students to their next class period. This year is going to be a 180 from last year…

Two weeks later, I witness a sponge fly across the room. That's right, a sponge. (Reread Chapter 2 on working with a mentor for a refresher on the original sponge debacle.)

It's déjà vu from my first year. I have found myself making some of the same mistakes as I did last year *and* a whole host of new ones. For example, calling a student by their sibling's name, running off the wrong number of handouts based on my old class numbers, and even handing out a disclosure that has the same typos from my first year including the wrong academic year. I feel like my "veteran" self-descriptor is already in jeopardy and I am only a few weeks into the second school year. All I can think is "But I'm not a new teacher anymore … am I?"

THE MORAL

Let's not drag this out: The answer was clearly "YES, you are still a *new* teacher." As much as I wanted to emerge from the struggles of my first year as a conquering hero, I was far from finishing this marathon of learning and experience that is required of all educators. There is a reason that statistics on new educators refer to teachers who are in their first *five* years of the profession. It takes multiple years to develop and hone your skills with students in the classroom.

Before you are daunted by the number of years, let me say that you have already been through the most difficult part of this journey by completing your first year. The degree of difficulty here is similar to a 1/x function (ask your physics teacher friends if that doesn't ring a bell) so you have already conquered the steepest part of the challenge. But just because you have climbed that hill doesn't mean you are at the top of the mountain, as I thought I was at the outset of my second year. So to prepare yourself for the continuation of your journey, here are some tips to help you earn that "veteran" title sooner rather than later.

STEPS FOR SUCCESS

1. KEEP A RECORD OF EDITS, THOUGHTS, AND SUGGESTIONS FOR THE FUTURE.

Here is where computers are worth their weight in gold. As you implement lessons, labs, and units, keep a master plan that you can quickly make notes, additions, and changes based on your classroom experiences. This needs to be easy and accessible, so I recommend a digital document if at all possible. This document can then serve as your planning guide for the following year. As teachers, we all like to think that we have memories like a steel trap and will remember that potassium chlorate was the best reactant to use in the chemistry demonstration, but after a year of chemicals, dissections, engineering cars, and answering adolescent questions, the cobwebs can start to grow. By keeping a record, you can make immediate changes … or changes in three weeks when you have a chance to catch back up.

2. MAKE THE CHANGES YOU SUGGEST.

This seems very obvious based on the step suggested above. However, it can be difficult to find the time to update plans for the future while taking care of your day-to-day responsibilities as a teacher. However, this investment is invaluable in supporting your continued improvement as an educator. Set aside 5–10 minutes at the end of the day to make your suggested changes if they are quick. If it's a complete rewrite (which does happen beyond your first year of teaching), take a moment to document what went well and what you would like to do differently in general terms. This will give you a heads-up the following year and prevent you from repeating the same experience that didn't work for you or your students.

3. SAVE, SAVE, AND SAVE AGAIN.

In this day and age of digital documents, it's not surprising to hear that most educators keep the majority of their school files on their computers. It is equally not surprising to hear horror stories about teachers having their computer crash and losing their entire curriculum in a single moment. This is not an urban legend, and reminds all teachers to be diligent about saving their curriculum in more than one location.

Aside from saving to your computer, I would recommend that you save additional copies to any or all of the following locations:

- *Your school server*. Many schools operate a network that can be used to back up files. These can be helpful as you can also share files within the school community, although access can be limited to devices that are hooked into the school network.

- *The Cloud*. Whether it's DropBox, Google Drive, or any of the others available, copying files to an online resource is another good backup step that alleviates any issues that may arise on a school server. This also gives you access to your files from any place that has Wi-Fi, which can be helpful if you want to share something immediately with another teacher you meet at a conference.

- *Hardware.* Having an external hard drive or even a large memory USB drive handy is another good option. This is another extra precaution in case the internet goes away (in which case, as educators we would have some additional issues to conquer). They are also very useful for sharing and accessing files when Wi-Fi isn't readily available.

The message here is that there are multiple options, but they are only as good as your diligence in saving and backing up. A USB drive makes a nice addition to the key ring, but isn't very helpful if the last time you updated it was over a year ago. There are multiple programs and devices that can help you keep track of this, and the best approach is to talk with your school IT personnel to see if they have suggestions or a school licensed program that they recommend. When in doubt, *always be saving*.

4. SEEK STRUCTURED STUDENT FEEDBACK THROUGHOUT THE YEAR.

Here is your moment to turn the tables and have your students support your learning. They are savvy consumers and are engaged in classes eight hours a day, five days a week. Students are an excellent resource for seeking feedback about your educational community and can help you gain insights about your instruction that you and your colleagues may not readily see. When setting up student feedback:

- *Anonymous is oftern best.* I always gave students the opportunity to share their names in case they had a specific concern and they wanted me to follow up specifically with them. However, most students opted for anonymous. Online survey systems such as Google Forms or Survey Monkey can be a great tool for maintaining confidentiality and getting honest feedback.

- *Structure your format to give you helpful feedback.* Many students are not familiar with the idea of "constructive criticism" versus "criticism." If you ask as student "What didn't you like about this unit?" be prepared to get the good, the bad, the ugly, and maybe even some rotten vegetables thrown your way. Instead, ask students "If you could change one lab, which one would it be and how would you change it?" It's also helpful to ask about their most enjoyable and enriching experiences. It was a good indicator for seeing which lessons they responded to, and look at those characteristics for updating the lessons that didn't rank near the top.

- *Ask for feedback periodically, adjusting your requests based on classroom climate and changes in topics.* If you only do this once or twice a year, the students don't get to see you implement the changes based on their feedback. If you open the lines of communication more regularly it will help build a culture of openness and trust with your students. It also gives your more data to help instruct the changes in your science classroom.

- *Share your feedback with your mentor(s).* Again, this can be a great conversation starter with colleagues to help you brainstorm improvements, solutions, and even help them reconsider some of their approaches. It fosters the strong communication that is encouraged in the chapter on working with a mentor.

5. KEEP IN TOUCH WITH YOUR MENTOR.

Many organizations only assign a mentor for first-year teachers, leaving second-year teachers to try and move into the "all-knowing veteran" trap. Although you may be fluent in the bell schedule, your mentor still has vast knowledge that can support you as you continue into your next few years of teaching. Keeping a set appointment or reminder on your calendar to check in can give you moments to share and seek community support that is still vital to your growth and development.

WHAT DOES SUCCESS LOOK LIKE?

"Did you really mean Phun with Physics, or is that another typo?" asks a student as I hand out an exam review activity. I'm not sure what was worse, that I thought my title was clever or that I was known for typos amongst my students. (To be fair, many of these lessons had come from a Google search the night before, and my grammar and punctuation weren't always the best at 1:00 a.m. Yes, even as a second-year teacher, I was still spending late nights planning.)

"Great! I'm glad you caught that! Now see if you can find the other four that I purposely included to help you improve your observation skills," I respond with a smirk. Although I hadn't finessed all of my assignments to perfection as a second-year teacher, at least I had improved my witty responses to students.

I can say that flying sponges occurred well into my fourth year of teaching before a teacher from California I met at a conference recommended that I keep a teacher bucket of sponges that I only handed out during clean up. (Thank you, teacher in California.) The important thing was that I took time to acknowledge my successes with students, and use those moments to inspire my continued learning and efforts in the classroom. Each day built experience, and that experience helped to create a foundation for my future in education.

RESOURCES FOR MORE INFORMATION

New Science Teacher Academy with the National Science Teachers Association (*www.nsta.org/academy*)

The NSTA New Science Teacher Academy, cofounded by the Amgen Foundation, is a professional development initiative created to help promote quality science teaching, enhance teacher confidence and classroom excellence, and improve teacher content knowledge.

According to a 2003 study by Richard Ingersoll, nearly 50% of beginning teachers leave their jobs in the first five years. The NSTA New Science Teacher Academy endeavors to use mentoring and other professional development resources to support science teachers during the often challenging initial teaching years and to help them stay in the profession.

Froschauer, L., and M. L. Bigelow. 2012. *Rise and shine: A practical guide for the beginning science teacher*. Arlington, VA: NSTA Press.

Rise and Shine provides a friendly support system that new science teachers can turn to in their first days, months, and even years in the classroom. This easy-to-read book offers plenty of helpful techniques for managing the classroom, maintaining discipline, and working with parents. But it also covers important topics unique to science teaching, such as setting up a laboratory, keeping the classroom safe, and initiating inquiry from the first day. Sprinkled throughout the book is candid advice from seasoned science teachers who offer both useful strategies and warm reassurance. *Rise and Shine* is designed to help preservice teachers, those in the first few years of teaching (regardless of grade level), and those who may be entering a new situation within the teaching field. If you need a mentor—or if you are a mentor or instructor who wants to support beginning science teachers—this book is for you.

McCourt, F. 2006. *Teacher man: A memoir*. New York: Scribner.

Now, here at last, is McCourt's long-awaited book about how his 30-year teaching career shaped his second act as a writer. *Teacher Man* is also an urgent tribute to teachers everywhere. In bold and spirited prose featuring his irreverent wit and heartbreaking honesty, McCourt records the trials, triumphs, and surprises he faces in public high schools around New York City. His methods anything but conventional, McCourt creates a lasting impact on his students through imaginative assignments (he instructs one class to write "an excuse note from Adam or Eve to God"), singalongs (featuring recipe ingredients as lyrics), and field trips (imagine taking 29 rowdy girls to a movie in Times Square!).

McCourt struggles to find his way in the classroom and spends his evenings drinking with writers and dreaming of one day putting his own story to paper. *Teacher Man* shows McCourt developing his unparalleled ability to tell a great story as, five days a week, five periods per day, he works to gain the attention and respect of unruly, hormonally charged or indifferent adolescents. McCourt's rocky marriage, his failed attempt to get a PhD at Trinity College, Dublin, and his repeated firings due to his propensity to talk back to his superiors ironically lead him to New York's most prestigious school, Stuyvesant High School, where he finally finds a place and a voice. "Doggedness," he says, is "not as glamorous as ambition or talent or intellect or charm, but still the one thing that got me through the days and nights."

Codell, E. R. 2009. *Educating Esmé: Diary of a teacher's first year.* Chapel Hill, NC: Algonquin Books.

A must-read for parents, new teachers, and classroom veterans, *Educating Esmé* is the exuberant diary of Esmé Raji Codell's first year teaching in a Chicago public school. Fresh-mouthed and free-spirited, the irrepressible Madame Esmé—as she prefers to be called—does the cha-cha during multiplication tables, roller-skates down the hallways, and puts on rousing performances with at-risk students in the library. Her diary opens a window into a real-life classroom from a teacher's perspective. While battling bureaucrats, gang members, abusive parents, and her own insecurities, this gifted young woman reveals what it takes to be an exceptional teacher.

Heroine to thousands of parents and educators, Esmé now shares more of her ingenious and yet down-to-earth approaches to the classroom in a supplementary guide to help new teachers hit the ground running. As relevant and iconoclastic as when it was first published, *Educating Esmé* is a classic, as is Madame Esmé herself.

Palmer, P. J. 2007. *The courage to teach: Exploring the inner landscape of a teacher's life.* San Francisco, CA: Jossey-Bass.

"This book is for teachers who have good days and bad—and whose bad days bring the suffering that comes only from something one loves. It is for teachers who refuse to harden their hearts, because they love learners, learning, and the teaching life." — Parker J. Palmer [from the introduction]

For many years, Parker Palmer has worked on behalf of teachers and others who choose their vocations for reasons of the heart but may lose heart because of the troubled, sometimes toxic systems in which they work. Hundreds of thousands of readers have benefited from his approach in *The Courage to Teach*, which takes teachers on an inner journey toward reconnecting with themselves, their students, their colleagues, and their vocations, and reclaiming their passion for one of the most challenging and important of human endeavors.

This book builds on a simple premise: Good teaching cannot be reduced to technique but is rooted in the identity and integrity of the teacher. Good teaching takes myriad forms but good teachers share one trait: They are authentically present in the classroom, in community with their students and their subject. They possess "a capacity for connectedness" and are able to weave a complex web of connections between themselves, their subjects, and their students, helping their students weave a world for themselves. The connections made by good teachers are held not in their methods but in their hearts—the place where intellect, emotion, spirit, and will converge in the human self—supported by the community that emerges among us when we choose to live authentic lives.

LEADERSHIP FOR NEW TEACHERS

HOW TO TURN THE JOB INTO A CAREER

F inally, this chapter talks about how to translate your enthusiasm and energy into becoming a young leader in the world of science education. It will look at possible future career aspirations that science teachers can have, and how to achieve those by taking advantage of the ideas proposed in the previous chapters.

When looking at this final chapter, I had to ask myself, "How did I end up here?" It was not so long ago that I had taken the reigns of my first classroom and had jumped into the world of science education. I now have the great opportunity to work with the science education community as the K–12 Science Specialist with the Utah State Office of Education.

So how did I get from there to here?

THE STORY

I'm a third-grade student in Ms. Doyle's class at Homestead Elementary in Englewood, Colorado. It's time for science, and she has handed each student a lima bean. "Today we are going to learn about plants, and it all starts with a seed," she says as we start our investigation on photosynthesis.

It's the traditional "bean-in-a-bag" experiment, where you take a lima bean and place it between two wet paper towels and seal it in a plastic bag. You then tape the bags in the window and you wait.

With the right conditions, after a few days you see a small root emerge, followed by a small stem. Generally about half of the students' beans have either not sprouted or started to grow mold, but I was in the 50% of the students who were lucky enough to have their bean sprout!

At this point, we planted our seeds in small Styrofoam cups with soil, and continued to water and tend to them as the seedlings grew. Again, there is another round of mortality, but my bean plant continues to grow and thrive in the classroom. I am so incredibly proud of my plant that I bring it home at the conclusion of the experience, and plant it in a plastic bucket.

THE MORAL

"What's your favorite memory from your K–12 science education?"

This is a question I ask teachers, principals, and community members at the start of every presentation I do. As people share about projects, experiments, dissections, and field trips, I always come back to my bean-in-a-bag experience. It's the first time that I remember having a real "science" lesson where I had the chance to ask questions, engage in hands-on exploration, and learn in a way that created a memory to last a lifetime.

This is why I got into teaching. To be able to provide students with the "bean-in-a-bag experience" that they could point back to and say, "That is science." Whether it was on my first day of student teaching, my first lesson in my own classroom, or in my current role as a science specialist, I want kids to have their own opportunities to find their passion and appreciation for science.

Most science teachers have a specific teacher or a lesson that they attribute for their passion in science. If you haven't thought of yours yet, this is the moment to do so. Having a specific memory to use as a model for the type of science education that you want to provide to your students helps to create a tangible experience that you can use as your measure for the education you provide in your classroom.

The memory also helps you to set the bar high for what you expect from yourself as an educator. In the same way we ask students to go beyond their comfort zone to grow and improve in knowledge and skill, you need to mirror that same advice in your career as a teacher. By never settling and constantly pushing yourself, you will become a better science teacher year after year. This will, in turn, translate into a better science education for your students.

Striving to provide the "bean-in-a-bag" has been the catalyst for the experiences that not only helped me to develop my skills as an educator to survive my first few years, but it also has opened doors to leadership opportunities. I did not begin my first year of teaching with plans to be where I am today. I can say, however, that it was an investment in providing the best science education I could for students that led me where I am.

You are already demonstrating commitment by taking the time to seek out resources to support your efforts in the classroom. Even as a new teacher, you can continue down this path to pursue leadership opportunities within your school, your district, your state, and even the nation.

STEPS FOR SUCCESS

1. TAKE PRIDE IN YOUR CLASSROOM.

If you are interested in being a leader, the first step is to take pride in your classroom. Your students are the ones who are going to establish your reputation within the school community. By focusing your time on *their* learning and school experience, you will find that respect from your colleagues will follow without having to actively seek recognition. If you are invested in your students, the community will see your dedication and start to recognize your hard work and commitment.

2. GET INVOLVED IN YOUR COMMUNITY.

Engage with your students, parents, and colleagues beyond your science classroom. This approach is highlighted with more specific ideas in Chapter 9 about coaching and community, but it's a great way to establish relationships and show your investment in the community. It's also an excellent way to get to know what resources exist for improving science for your students, your science content knowledge, and your skills as an educator.

3. VOLUNTEER.

Let your lead teacher or department chair know that you are interested in helping organize, oversee, or even create science opportunities. There are always projects and committees that are taking place where you can provide a fresh perspective of a young educator. Joining these experiences is helpful for getting to know how a school operates, which is essential knowledge for all teachers who want to become leaders.

It's important that you voice interest in participating in these experiences because many times new educators are overlooked for these roles. Although sometimes it can be based on less experience, most of the time it's because those in charge don't want to overburden you in your first few years of teaching. I recommend finding one committee or project that is a specific area of interest to you. Once you have spent a year in that role, look at whether or not you want to volunteer for additional experiences. This can help to ease you into the experience without over-committing.

Finally, it's important to note that these experiences are generally a time investment that doesn't have a financial payoff. You have to keep in mind that the knowledge you gain is worth your time and effort when taking on these roles.

4. PURSUE PROFESSIONAL DEVELOPMENT.

In your professional life, you should model the learning and risks that you ask of your students. The importance of this step for your classroom is detailed in Chapter 10 on professional development (PD), and it's worth noting that professional development is equally important for future leadership opportunities.

Many teachers wait for these PD opportunities to find them, and they consider participation in school sponsored events as being adequate enough to fulfill this requirement. Don't let PD stop with "required" events. Look for those that are of interest to you, like a course on technology in the classroom with a local college or university. Like homework, there is little investment if you *have* to do it. It's much better if you can choose what courses make sense to you so you have an increased investment in the learning.

5. SHARE YOUR KNOWLEDGE.

You can be a teacher beyond just your K–12 students. It's a great experience to start sharing in your team meetings or professional learning community (PLC) or even at your next scheduled PD day. You can also consider presenting at the district level, or through a state science teacher's association.

And just like your first day of teaching, you will get better at presenting for other teachers the more that you practice doing so.

Recommendations for first-time presenters:

- Present with someone else. This helps to relieve some of the stress that comes with your first presentation.

- Have handouts or links to resources. Teachers appreciate being able to take ideas and move toward implementation without re-creating all of the tangible elements.

- Offer several variations of how to present the lesson or skill. It helps other teachers see how your suggestions would work best in their classrooms. This includes variety in equipment, grade-level connections, and assessment.

- Run too short? Give teachers time to ask questions and share their own ideas. Teachers like to talk, so feel free to give them the opportunity to share.

- Ask for feedback. It's important to get feedback from your peers so you can improve on your next presentation.

If you are *really* worried and you are about to start your initial presentation, tell the group this is your first time presenting. Your audience is often much more supportive if they know this, and it alleviates the pressure of being perfect.

Another opportunity to consider is to share your knowledge through written venues such as journals, newspapers, or even websites. These are great outlets for teachers who want to reach a wide audience without leaving home.

6. APPLY FOR OPPORTUNITIES.

Put yourself out there for experiences, even if you think they are beyond your reach. You never know when you might be at the right place at the right time. Many times, your interest and desire can translate into alternative opportunities. Besides, failure is good for everyone. It keeps us grounded, and can act as a challenge for future improvement. Examples include:

- Scholarships and grants for conferences, education, classroom materials
- Awards and fellowships
- Employment

7. NETWORK.

The field of education doesn't always do a great job of making programs, opportunities, and employment well known to a wide audience. It's your job as a teacher to build those communication links so you can have access to the knowledge that will help you achieve your goals. Business cards are a *great* way to keep in touch (yes, as a teacher, you can have a business card) with all the new colleagues you will meet at the various conferences you attend.

WHAT DOES SUCCESS LOOK LIKE?

My story begins with Chapter 1 of this book, and appropriately ends with this final chapter. My time in the world of education started with me standing in your shoes. I was a young teacher who had been trained by great professors and teachers. Their words of wisdom and advice gave me the foundation to venture into the classroom with excitement and promise.

Despite my excellent training, my first year was far from amazing. There were doubts, tears, and many lessons that I wish I could redo. But each challenge and failure inspired me to wake up the next day and do it better. By engaging with colleagues, continued learning, and self-reflection, I was able to improve and grow as a teacher.

I still strive to provide the "bean-in-a-bag experience" for students in science, and challenge you to do the same. There is no greater goal for a science educator, and it will help guide you to achieve your desired outcome.

RESOURCES FOR MORE INFORMATION

Albert Einstein Distinguished Educator Fellowship (*www.trianglecoalition.org/einstein-fellows*)

The Albert Einstein Distinguished Educator Fellowship Program offers current; public or private; elementary and secondary; science, technology, engineering, and mathematics classroom teachers with demonstrated excellence in teaching an opportunity to serve in the national public policy arena. Fellows provide practical insight in establishing and operating education programs. Fellowships increase understanding, communication, and cooperation between legislative and executive branches and the science, mathematics, and technology education community.

Albert Einstein Fellows bring to Congress and appropriate branches of the federal government the extensive knowledge and experience of classroom teachers. They provide practical insights and "real world" perspectives to policy makers and program managers developing or managing educational programs.

Distinguished Fullbright Awards in Teaching Program (*www.fulbrightteacherexchange.org/distinguished-fulbright-awards-in-teaching-program*)

Under the Distinguished Fulbright Awards in Teaching Program, highly accomplished U.S. primary and secondary level teachers of all subjects, guidance counselors, curriculum specialists, curriculum heads, talented and gifted coordinators, special education coordinators, and media specialists/librarians may apply for this professional development program to carry out a course of study for three to six months abroad.

The program provides U.S. teachers with the opportunity to study in an overseas research center or university and work within local schools in the host country. Applicants will propose a Capstone Project at the time of application that should enhance their learning and have practical applications to their teaching (see the Capstone Project page for sample projects).

Tweed, A. 2009. *Designing effective science instruction: What works in science classrooms.* Arlington, VA: NSTA Press.

Science teachers, like all teachers, start each school year with high hopes and expectations for students to succeed. They plan their lessons, scramble to get the necessary equipment, and work hard to engage their students. However, despite good intentions and best-laid plans, not all students do well in science classes, and even fewer achieve mastery. Student performance on national and international assessments is poor, and/or more adults are unable to understand the scientific issues that affect their lives and society. Something must be done now to help science teachers put power behind their hopes and expectations for student achievement.

Designing Effective Science Instruction helps you reflect on what is working well with your current approach to designing lessons and provides recommendations for improving existing lessons or creating effective new ones, all while exploring the characteristics of high-quality science lessons. Whether you are a novice or veteran teacher, the self-assessments and suggestions in this book offer guidance that encourages you to refine what you do to become a more effective science teacher.

Gess-Newsome, J., J. A. Luft, and R. Bell. 2008. *Reforming secondary science instruction.* Arlington, VA: NSTA Press.

Science education reform can seem a daunting task to high school science teachers. So, you might ask, why should I be bothered? The answer is that today's students simply do not have the skill sets necessary for life in our global economy.

Reforming Secondary Science Instruction offers detailed advice for changing your methods of teaching so that students are prepared for life and work. Follow along as your fellow teachers learn about inquiry, implement change strategies, try out innovative instructional materials, build professional learning communities and partnerships, use data from student assessments, and address the needs of linguistically diverse learners.

The underlying message, as one author puts it, is "science education reform will not occur by simply adding occasional new activities to your teaching repertoire. Reform requires thought, work, and persistence."

Every chapter offers you the opportunity to assess your own teaching techniques and find room for improvement. Whether you are early in your career or a seasoned professional, *Reforming Secondary Science Instruction* will help you craft a workable plan for giving your students the tools they need to succeed beyond your classroom.

INDEX